Innovations in GIS 3

*Selected papers from the
Third National Conference
on GIS Research UK (GISRUK)*

Innovations in GIS 3

Selected papers from the
Third National Conference
on GIS Research UK (GISRUK)

EDITED BY

DAVID PARKER

Department of Surveying,
University of Newcastle upon Tyne

Taylor & Francis
Publishers since 1798

UK   Taylor & Francis Ltd, One Gunpowder Square, London EC4A 3DE

USA   Taylor & Francis Inc., 1900 Frost Road, Suite 101, Bristol, PA 19007

Copyright © Taylor & Francis Ltd 1996

British Library Cataloguing in Publication Data

A catalogue record for this book is available from the British Library

ISBN 07484 04589 (paperback)

ISBN 07484 04597 (cased)

Library of Congress Cataloging in Publication data are available

Cover design by Hybert Design & Type, Waltham St Lawrence, Berkshire

Typeset in Times 10/12pt by Santype International Limited, Netherhampton Road, Salisbury, Wiltshire

Printed in Great Britain by T. J. Press (Padstow) Ltd

Contents

Foreword

Having witnessed the third GISRUK conference and being impressed with the breadth and quality of material on show, it is a reasonable time to reflect on the maturation of GIS research. Several years ago I wrote a piece for *Environment and Planning A* entitled 'What does "Doing a PhD in GIS" mean?' in which I expressed the view that it was rather difficult to undertake a PhD in GIS and indeed equated it with undertaking a PhD in word-processing: both being software designed as a means to an end rather than as ends in themselves. The piece was prompted both by a deadline from the journal and by the rash of postgraduate students I had seen in the US who were convinced that all it took to obtain a PhD in geography was knowledge of the appropriate macro language in whatever GIS they were using. Many of these students found great difficulty in finishing their degree because they were too focused on the means and not enough on the ends.

I carry these thoughts with me to GIS conferences and often judge GIS-related papers by the amount of 'value-added' to the research by the GIS. I consider papers to have 'failed' this examination, for example, if they merely display their results on a map (probably not a very good one) created by a GIS when such results could have been displayed with greater clarity using some standard mapping package. Similarly, I judge a paper to have failed if it presents some standard analytical tool transported to a GIS environment. A necessary, but not sufficient, condition for a good paper at a GIS conference should be a description of how the GIS has prompted the author to do something that probably would not have been done otherwise. Notice that I do not say '*could* not be done without a GIS' as this is much too restrictive and we would have conferences virtually devoid of papers!

Returning to the theme of the maturation of GIS research, I believe that many of the papers presented at the early GIS conferences would not have passed my test; however, perhaps this was a necessary stage in the development of the subject matter when we were just coming to grips with the powder of GIS. More recently, and certainly within this volume, I have seen an increasing number of papers which demonstrate the 'value-added' of GIS in different ways and which signal the maturation of research in this field. These papers tend to provide 'value-added' in different ways:

1. Care and preparation of data. The use of GIS has prompted us to ask basic questions about the accuracy and reliability of our data, about error propagation and about the sensitivity of our results to the way in which our data are collected. These questions in turn have prompted basic research on issues that would otherwise have been neglected.
2. The representation of topology and spatial relationships. How topology can best be represented and how can we utilise this information in spatial models and spatial analytical procedures to maximum advantage?
3. The utilisation of the display capabilities of a GIS. This is prompting new ways of looking at data with statistical displays linked to maps, with data draped over surfaces and with the development of local or 'mappable' statistics as opposed to the more traditional global or 'whole map' statistics.

This then is where the challenge of GISRUK lies – to act as a forum for papers that demonstrate the value-added nature of GIS research and at the same time add to the value of GIS as an important tool in the analysis of spatial data. Collections of papers such as this confirm the impression that not only has GIS re-awakened interest in issues of spatial data across many disciplines but also the level of research is maturing rather rapidly.

A. STEWART FOTHERINGHAM

Professor of Quantitative Geography
Director, NERRL
September, 1995

Preface

The chapters in this volume are enhanced and revised versions of papers presented and discussed at the annual UK National Conference on GIS Research: GISRUK'95. The conference was held at the University of Newcastle upon Tyne in the North East of England in the Spring of 1995.

The origins of the conference lie with a small group of United Kingdom researchers from a range of disciplines but all with interests in GIS. In early discussions some 4 years ago it was felt that there was the need for a new forum for UK GIS research. The idea of a conference was formed, not one to compete with the existing national meetings and not one to compete with existing international research conferences. The aims of the new conference were to:

- act as a focus for GIS research in the UK
- provide a mechanism for the announcement and publication of GIS research
- act as an interdisciplinary forum for discussion for research ideas
- promote active collaboration among UK researchers from diverse parent disciplines
- provide a framework in which postgraduate students can see their work in a national context.

The conference this past spring was the third in the series. As well as sessions for the formal presentation and discussion of research results, there were other events such as a debate, round-table discussions and poster displays. The mix of event types was all aimed at achieving the objectives of fostering exchange of ideas and collaboration in an informal and friendly atmosphere and all at a price to encourage researchers, no matter what their status.

This book is the third in the *Innovations* series, the last two being the edited proceedings of the first two conferences. The series is designed to satisfy the second of the aims set out above, that of providing a mechanism for the publication of GIS research. There are 18 chapters in this book. Two of them are by the keynote speakers, Leila De Floriani and Michael Batty. The other 16 have been selected from over 80 contributions originally offered to the conference. Extended abstracts were first refereed by the programme committee formed from both the steering committee and the local organising committee. The full versions of the papers were then presented and discussed at the conference. The raw published chapters were selected by a sub-set of the programme committee after the conference and the authors given the opportunity to revise their text before final acceptance for *Innovations 3*.

This preface gives me an opportunity to thank the many people and organisations who have been involved with the GISRUK conferences and the *Innovations* series. The major sponsors of GISRUK'95 were the UK Association for Geographic Information (AGI). They have sponsored the event from its conception through their Education and Information Committee and without them this research forum would not have achieved the success it has. Thanks are also due to the other long-term sponsors, the Regional Research Network (RRLNET). This organisation has recently gone through a metamorphosis but through it all they continued their support. For the first time in 1995 we invited publishers' representatives to display their texts and other media for the GIS education and research market. This proved successful both in informing the delegates as to current publications and in raising further sponsorship. It is hoped that the publishers found their input beneficial to themselves and will continue to support the event.

My thanks go to all the steering committee and local organising committee members. Their names are listed elsewhere in the text. One must not underestimate the amount of time and effort in soliciting, reviewing and processing offered papers, in attracting, registering and marshalling delegates and in raising finance to fund it all. Thank you all. I must select out two people for a special mention. Mike Worboys has just stepped down, having chaired the steering group over the past 3 years. It was Mike who drew the group together in the first place. It was Mike who lead the first local organising committee for the conference at Keele. It was Mike who negotiated the arrangements for and edited the first text in the *Innovations* series. Mike has just taken up a chair in Computing Science at Keele University. On behalf of all the steering committee I offer him our thanks and best wishes for the future. The second person to single out is Jan Rigby who, throughout the organisation of the 1995 conference and this book, never failed us with her calm, pragmatic and efficient administration always tempered with good humour. I must not forget to mention the contributors to both the conference and the book. With them there would be nothing for us to organise. My thanks to all.

DAVID PARKER
Newcastle upon Tyne, 1996

Contributors

Alia I. Abdelmoty
University of Glamorgan, Department of Computer Studies, Pontypridd,
Mid Glamorgan, Wales CF37 1DL
(aiabdel@glam.ac.uk)

Moh'd B. Al-Daoud
School of Computer Studies, University of Leeds, Leeds LS2 9JT
(al@scs.leeds.ac.uk)

Tigran Andjelic
Department of Computer Science, Keele University, Staffordshire ST5 5BG
(tigran@uk.ac.keele.cs)

Richard J. Aspinall
Macaulay Land Use Research Institute, Craigiebuckler, Aberdeen AB9 2QJ
(r.aspinall@mluri.sari.ac.uk)

Michael Batty
Centre for Advanced Spatial Analysis, University College London, Gower Street,
London WC1E 6BT
(mbatty@geog.ucl.ac.uk)

Chris Brunsdon
Department of Town & Country Planning, University of Newcastle upon Tyne,
Newcastle upon Tyne NE1 7RU
(chris.brunsdon@ncl.ac.uk)

Hugh Buchanan
Department of Surveying, University of Newcastle upon Tyne, Newcastle upon Tyne
NE1 7RU
(hugh.buchanan@ncl.ac.uk)

Martin Charlton
Department of Geography, University of Newcastle upon Tyne, Newcastle upon Tyne
NE1 7RU
(martin.charlton@ncl.ac.uk)

Leila De Floriani
Department of Computer and Information Sciences, University of Genova, Via Dodecaneso 45, 3–16146 Genova, Italy
(deflo@disi.unige.it)

Christine E. Dunn
Department of Geography, University of Durham, Science Site, South Road, Durham DH1 3LE
(c.e.dunn@durham.ac.uk)

Jason Dykes
Department of Geography, University of Leicester, Leicester LE1 7RH
(jad7@le.ac.uk)

Alistair Geddes
North West Regional Research Laboratory, Department of Geography, Lancaster University, Lancaster LA1 4YB
(a.geddes@lancs.ac.uk)

Gary Higgs
Department of City and Regional Planning, Cardiff University of Wales, PO Box 906, Cardiff CF1 3YN
(higgsg@uk.ac.cf)

Mike H. W. Hobbs
Department of Computer Science, University of Essex, Colchester CO4 3SQ
(mikeh@uk.ac.essex)

Crispin Hoult
Department of Surveying, University of Newcastle upon Tyne, Newcastle upon Tyne NE1 7RU
(c.hoult@ncl.ac.uk)

David Howes
National Center for Geographic Information & Analysis, State University of New York, 301 Wilkeson Quad, Buffalo, NY 14261-0023
(howes@zia.geog.buffalo.edu)

Simon P. Kingham
Institute of Environmental & Policy Analysis, University of Huddersfield, St Peter's Building, St Peter's Street, Huddersfield HD1 1RA
(s.p.kingham@hud.ac.uk)

Andrew G. Larner
School of Surveying, University of East London, Longbridge Road, Dagenham, Essex RM8 2AS
(larner@uk.ac.uel)

Allan Lilly
Macaulay Land Use Research Institute, Craigiebuckler, Aberdeen AB9 2QJ
(a.lilly@mluri.sari.ac.uk)

Robert MacFarlane
Division of Geography and Environmental Management, Lipman Building, University of Northumbria, Newcastle upon Tyne NE1 8ST
(robert.macfarlane@unn.ac.uk)

David Martin
Department of Geography, University of Southampton, Southampton SO17 1BJ
(d.j.martin@uk.ac.soton)

Paola Marzano
Department of Computer and Information Sciences, University of Genova, Via Dodecaneso 45, 3-16146 Genova, Italy
(marzano@disi.unige.it)

David R. Miller
Macaulay Land Use Research Institute, Craigiebuckler, Aberdeen AB9 2QJ
(d.miller@mluri.sari.ac.uk)

Jane G. Morrice
Macaulay Land Use Research Institute, Craigiebuckler, Aberdeen AB9 2QJ
(j.morrice@mluri.sari.ac.uk)

Stan Openshaw
Centre for Computational Geography, School of Geography, University of Leeds, Leeds LS2 9JT
(s.openshaw@geog.leeds.ac.uk)

David Parker
Department of Surveying, University of Newcastle upon Tyne, Newcastle upon Tyne NE1 7RU
(david.parker@ncl.ac.uk)

Norman W. Paton
Department of Computer Science, University of Manchester, Oxford Road, Manchester ME13 9PL
(norm@cs.man.ac.uk)

Tim Perrée
Centre for Computational Geography, School of Geography, University of Leeds, Leeds LS2 9JT
(t.perree@geog.leeds.ac.uk)

Enrico Puppo
Institute for Applied Mathematics, National Research Council, Via De Marinie, 6-16149 Genova, Italy

Stuart A. Roberts
School of Computer Studies, University of Leeds, Leeds LS2 9JT
(sar@scs.leeds.ac.uk)

George Taylor
Department of Surveying, University of Newcastle upon Tyne, Newcastle upon Tyne NE1 7RU
(george.taylor@ncl.ac.uk)

M. Howard Williams
Department of Computing and Electrical Engineering, Heriot-Watt University, Riccarton, Edinburgh EH14 4AS
(howard@cee.hw.ac.uk)

Joseph Wood
Department of Geography, University of Leicester, Leicester LE1 7RH
(jwo@leicester.ac.uk)

Michael F. Worboys
Department of Computer Science, Keele University, Staffordshire ST5 5BG
(michael@uk.ac.keele.cs)

Gary G. Wright
Macaulay Land Use Research Institute, Craigiebuckler, Aberdeen AB9 2QJ
(g.wright@mluri.sari.ac.uk)

GISRUK Committees

GISRUK National Steering Committee

Richard Aspinall	MLURI, Aberdeen
Heather Campbell	University of Sheffield
Peter Fisher	University of Leicester
Bruce Gittings	University of Edinburgh
Zarine Kemp	University of Kent
David Parker	University of Newcastle upon Tyne
Jonathan Raper	Birkbeck College, University of London
Michael Worboys	Keele University

GISRUK'95 Local Organising Committee

David Parker
Chris Brunsdon
Hugh Buchanan
Chris Dunn
David Fairbairn
Stewart Fotheringham
Colin McClean
Jan Rigby
George Taylor

GISRUK'95 Sponsors

Association for Geographic Information
Regional Research Laboratory Network

Introduction

This book contains a collection of papers describing current research into geographical information systems (GIS). GIS is something of an unfortunate title as it does not suit all the disciplines that have an interest in the topic. Some disciplines prefer 'spatial' rather than geographical since they consider themselves unrelated to the discipline of geography. The word 'system' conjures up the image of a turnkey package of hardware and software and although these exist as tool boxes for applications, the field of GIS is much broader than this. Whatever the arguments, GIS has become accepted as the term to encompass the field of the input, modelling, management, retrieval, analysis and presentation of spatially referenced information. The range of disciplines that contribute to GIS is wide – geography, mathematics, computing science, surveying, archaeology, planning, medicine – and all of these contributed to the papers from which the book is drawn.

Contributions have been selected on the basis of their innovation to GIS as a subject and not for their advancement of a particular discipline by the application of GIS techniques. This is not always clear from the titles of the chapters from which it might appear that the content is discipline specific. Within all the chapters there is interest for all those wishing to learn more about current research in GIS.

Within the book the chapters have been divided into five sections, these relate quite closely to the stages listed in the rather open definition of GIS given above. The sections are entitled data issues, computational support, spatial analysis, visualisation and applications. The division between the sections is not strict. It proved impossible to divide the continuum of current GIS research activity under only five heads.

DATA ISSUES

The five chapters in this section span the currently hot topics of multi-resolution data modelling, data availability and data reliability.

Leila De Floriani's paper, written in collaboration with **Paola Marzano** and **Enrico Puppo** was one of the two invited contributions to the GISRUK'95 conference. Her theme was multiresolution modelling applied to the representation of terrain surfaces. However, the chapter does have relevance for wider issues of multi-resolution modelling in spatial information theory. The importance of the work is based on the premise that to support future generations of spatial information systems it is worthwhile to pursue the definition of a formal framework for multi-resolution representation of spatial entities based on a topological model that offers explicit description of spatial objects and efficient retrieval of spatial relations. In the chapter, the concept of a digital terrain model is formalised and the problems of terrain approximation considered. A formal definition of multiresolution modelling is introduced that allows a systematic description of all proposed structures. The current state of the art in multiresolution terrain modelling is then illustrated with a series of examples – all this with a view to future extensions.

The chapter by **David Parker, George Taylor** and **Crispin Hoult** takes a rather more pragmatic approach to multiresolution data modelling. The work is significant for environments like those in UK local authorities who will soon be facing the problems of managing spatial datasets where objects of the same class will need to be represented in different ways: a property represented by anything from a single point to a surface model with ground detail. The proposed model is based on the land face as the basic unit. This can be a practical unit in the UK where the Ordnance Survey basic scale data can be used as a base. The chapter includes a description of the implementation of the model for the surveyor as the person who is responsible for raw data supply. It is shown how face-based data can be very practical for the supply of spatial information to the wide variety of users with their wide variety of systems.

Andrew Larner's chapter arises out of his research into the legal and institutional restrictions upon the handling of land related data in digital form. This type of research is not typical of what has been presented to date in the Innovations series. It is essential however. If data supplier' rights are not fairly protected then other innovations in GIS will not prove beneficial as essential data will not be released. The chapter starts with a review of what is meant by rights and interests in data. It then examines what is meant by information and data and how the legal framework protects interests in data. A new method of protecting rights is proposed which is based upon examination of the full spectrum of current intellectual property and real property law as well as the history of copyright.

The next chapter, by **David Martin** and **Gary Higgs**, investigates the increasingly detailed georeferencing systems in Britain. The use of postcodes and census enumeration zones has for a long time provided the only widely used means of geo-referencing population and socio-economic data. However, many organisations have built up extensive databases of address referenced or property referenced information. In the chapter are illustrated the many different ways in which existing databases referenced in this way may be manipulated within standard GIS software in order to provide hybrid geographical references. The work highlights the importance of the widespread adoption of the new British Standard for address referencing.

David Miller, Richard Aspinall, Jane Morrice, Gary Wright and **Allan Lilly** have contributed a chapter investigating the assessment of catchment environmental characteristics and their uncertainly. This work has obvious interest for those in hydrological modelling but it is in this position in the text for its consideration of data issues. It is an example of the construction of a dataset maintaining more than the minimum information to enable a variety of uses and assessment of reliability.

Many hydrological models operate on catchments as an entity: their characteristics are simplified to a single entity and their variability ignored when considering reliability. This chapter reports on the issues and the lessons learnt have implications for a wide audience. Note the demonstration of variability of interpretation of results from different scales of mapping. The same catchments are interpreted from 1 : 10 000, 1 : 25 000 and 1 : 50 000 maps with interesting results.

COMPUTATIONAL SUPPORT

Under this title have been grouped the contributions concentrating on database issues and algorithms.

Tigran Andjelic and **Michael Worboys** tackle the problems of version management for GIS, concentrating on the problems raised in a distributed environment. In particular, they consider the issues raised when the GIS is being used for design activities. Here there are special requirements, including the management of more than one alternative plan and the handling of transactions that are more complex and lengthy than traditional database interactions. For design work, large tasks are often subdivided and each group of tasks assigned to a group of designers. Some tasks are even sub-contracted to external organisations. A particular problem lies in the need for 'parallelism' rather than a serialised sequence in transactions. These requirements are additional to standard GIS demands such as that for spatially seamless data. Their review of the traditional serialised and short transaction handling processes and the longer and version-managed processes lead them to test the implementation of integrating their suggested model that corresponds to the 'read-write partial interface enabling extension of the long transaction environment into the short transaction environment'. A transaction environment is shared between the version-managed GIS database of Smallworld and the short-transaction-managed external relational database of Oracle. This shared environment follows the copy-on-update principle.

Commercial relational databases have been found to be inadequate for supporting geographic applications because of their spartan data modelling facilities and the limited computational power of their query languages. This has lead to the use of coupled systems where the spatial data manager and the database are distinct components with consequent disadvantages for programmer productivity and run-time performance. This is the starting point for the work by **Norman Paton, Alia Abdelmoty** and **Howard Williams**. They identify the range of facilities which it would be desirable for a spatial database system to support. They compare a number of software architectures against this range of facilities. The three representative spatial database systems are considered: GRAL, ONTOS and ROCK & ROLL. It is seen that kernel support for spatial data types is likely to yield improved performance and that certain architectures provide a fixed set of spatial data types and operations that are difficult to extend. The work goes on to describe their experience in applying an object-oriented model OM, an imperative database programming language ROCK and a logic language for writing inference rules and expressing queries ROLL to spatial data management.

The contribution of **Mike Hobbs** is especially interesting for a couple of reasons. It gives a good insight into the potential for genetic algorithms for spatial problems. The main attraction is said to be the ability to apply a simple search technique, modelled on the biological processes of evolution and natural selection, to problems with vast numbers of potential solutions. Whilst describing the implementation the concepts of fitness functions and crossover and mutation functions are introduced. The application described here will be close to the heart of all those who have been through the trauma of house-hunting: preclassifing houses into different types depending on their location. The road network was used as the basic areal unit but it is claimed that others could be used. The results, or a few of the many possible solutions, are shown demonstrating problems with defining the size of the areas of classification. The significance of the work is claimed to be that the algorithm provides a spatial analysis tool that is robust and friendly to inexperienced users.

Mohammed Al-Daoud and **Stuart Roberts** in their contribution investigate the problems of the extensive use of the nearest neighbour algorithms as required for spatial analysis operations such as clustering. They review three methods: K-d trees, quadtrees and the cell method. For each they tabulate time requirements and drawbacks. K-d trees and quadtrees are found to be very similar in both these criteria. The cell method is found to have few drawbacks but constant time demands. To test suitability for clustering the quadtree and cell methods are implemented and compared for a range of uniform and non-uniform data. The cell method is then enhanced by optimising the number of searches of circles for cell content. The results show that time savings of up to 16 per cent can be achieved but decreases when the data are non-uniform.

The last chapter in the section on computational support is rather different in emphasis. It reports on the computational techniques for determining topology: not two-dimensional but three-dimensional topology. **Hugh Buchanan**, as well as approaching the subject avoiding extensive mathematical terminology, approaches it from the need to develop data-cleaning and validation tools rather than from the need for such techniques in the field of solid modelling. Validation processes such as these are becoming required in testing conformance of datasets with published transfer standards. The work includes a different computational approach to a boxing test used to optimise the testing procedures.

SPATIAL ANALYSIS

The two chapters in this section are drawn from those offered to the Regional Research Network sponsored session.

The chapter by **Stan Openshaw** and **Tim Perrée** outlines the development of a new approach to spatial analysis in GIS environments. They attempt to develop an abstract but intuitively obvious spatial analysis process that is designed to be readily understandable to the typical non-statistically minded end-user. The issues they face are the apparent complexity of current technology, its inability to provide useful results, the difficulty the end user has in understanding results and the need of the end user to communicate the meaning and significance of spatial analysis results to others. There are four components to their system design. An automated exploratory mechanism that searches spatial data for patterns, a measure of search result performance used to guide the search process, an interpretation module to highlight important information and finally a means of visualising the search process itself. The success of the new method is evaluated using four different simulated datasets that simulate a rare disease with a frequency similar to childhood leukaemia. The final product is a series of MPEG movies available for viewing on the World Wide Web.

Chris Brunsdon and **Martin Charlton** continue the search for better and more intuitive spatial analysis functionality in GIS systems. Like the Openshaw paper they point to claims that the facilities for management and manipulation of spatial data are well catered for but there is little provision for statistical modelling and the investigation of relationships. The current solution for providing better high-level statistical modelling is to interface GIS information with one of the many statistical packages. Interfacing may be as simple as transfer via a text format. A different approach is taken here, one which extends the data handling and mapping facilities of a package designed for statistical programming. The package developed and described here utilises the Lisp programming language and in particular its implementation in XLisp-Stat: a Lisp interpreter and with extra statistical functionality – and available in the public domain! The work describes the representation of geographical information using complex numbers for points in a two dimensional plane. It describes 'brushing' facilities where highlighting parts of a histogram with a mouse automatically highlights corresponding points on a scatter plot.

VISUALISATION

The three chapters are all quite different and discuss broader issues than just visualisation, but this is their main thrust.

Michael Batty was the second invited speaker at GISRUK'95. His contribution, co-authored by **David Howes**, explores urban development dynamics through visualisation and animation. New micro scale administrative data is becoming available for describing urban areas. Other physical data such as remotely sensed imagery is also available. This chapter is first concerned with assembling this variety of data for use in the analysis of urban development, and part of the work explores a sample of such data for the Buffalo metropolitan region. The chapter concentrates on the setting up procedures for visualisation and computer animation of urban space–time patterns. This work is still in progress and future steps are discussed.

The chapter by **Joseph Wood** is another one that might seem misplaced in the text. If there was a section on digital elevation modelling then his work on scale-based characterisation could equally well have been placed there. There have been improvements in automated DEM characterisation in recent years but there is still a failure to take into account the effect of scale on the morphometric properties of the surface. The research on which this chapter is based considers methods of morphometric characterisation that involve a multi-scale description of surface form: multi-scale quadratic approximation. The method is shown to be an improvement as a mechanism for terrain model smoothing as required for more realistic shaded relief and as a basis of terrain visualisation.

The chapter by **Jason Dykes** describes the use of graphical user interface widgets in computer cartography. The aim of the work is to enable the non-specialist cartographer to control model and display parameters and hence change their view of a set of data in real time. This recognises the fact that many datasets contain vast amounts of complex temporal and spatial information, too much and too varied to be presented in a static mode. The tools used are 'Tcl', a windowing programming environment, and 'Tk', an X11 toolkit which defines familiar widgets such as buttons. The tools are in the public domain and extensions are being constantly produced and so the potential for further development whilst minimising effort is good. The chapter details where on the World Wide Web the developed code can be accessed.

APPLICATIONS

These three contributions, as the section title suggests, discuss more the application of GIS rather the innovation of the subject itself. However, there is research in all these chapters of relevance to a wider audience than the discipline of the application.

In the first contributed chapter in the applications section, **Alistair Geddes** investigates one aspect of a GIS-based field information management system. The work concentrates on the creation of yield maps: the positioning of the harvester using GPS, the yield meter measurements with the problems of relating the measure to field position, the integration of the GPS position and yield and then the spatial interpolation of the yield maps themselves. A kriging method of interpolation was adopted. The discussion raises the broader issues of the acceptance and usefulness of GIS for farm-based management. Current users of yield mapping use the result as a visual guide only. There is some way to go before one is able to make proactive judgements based on analysis of crop performance and other information.

Relationships between environmental exposure and ill-health are highly complex. **Christine Dunn** and **Simon Kingham** in their chapter are investigating a GIS framework to help to achieve a full understanding of the processes and mechanisms involved. The work has wide interest to the GIS community as an example combining specialist modelling with more traditional GIS techniques. They use air quality models in an attempt to replace suitable measured data. In their tests on a case study they show no positive associations between proximity to a source of pollutants and health survey data. There may not be, of course, but their discussion highlights the problems: how exposure to a toxin can vary with lifestyle, how exposure and response are related, the reliability of the air quality model used.

The final contribution to the applications section has again been selected for its potential wide interest to the GIS community. It questions the validity of only pursuing applications of GIS for which there are adequate data where human behaviour is known to be a very significant factor in the understanding and accuracy representation of change. The study described by **Robert MacFarlane** concentrates on land-user intentions and land-use modelling. The research has developed a GIS-based tool to assess the implications of various policy changes on specific areas of land. It shows that it is possible to construct complex models of landowner and land-user behaviour which will yield an accurate picture of the consequences of shifts in decision makers' external environments.

PART ONE

Data Issues

CHAPTER ONE

Multiresolution modelling in geographical information systems

LEILA DE FLORIANI, PAOLA MARZANO and ENRICO PUPPO

1.1 INTRODUCTION

The representation of spatial data at different resolutions is a topic of relevant interest in spatial information theory. Multiresolution modelling offers interesting capabilities for spatial representation and reasoning: from support to map generalization and automated cartography, to efficient browsing over large GIS, to structured solutions in wayfinding and planning (Frank and Timpf, 1994). Current GIS do not offer much in multiresolution data handling: apart from some hierarchical capabilities in raster modelling, which are essentially based on structures and tools inherited from image processing, there is an almost total lack of features for handling spatial data at different resolutions. In order to support future generation GIS, it seems worthwhile to pursue the definition of a formal framework for multiresolution representation of spatial entities based on a topological model that offers explicit description of spatial objects, and efficient encoding/retrieval of spatial relations. Some research has been undertaken in the last few years on the development of data models for the multiresolution representation of maps in the context of GIS (Bruegger and Frank, 1989; Bruegger and Kuhn, 1991; Bertolotto et al., 1994c; Frank and Timpf, 1994; Puppo and Dettori, 1995). However, such studies can still be considered at their initial stages. A considerable amount of work has been done on the development of multiresolution models for representation, manipulation, and visualization of terrains. Terrain models are obtained on the basis of a discrete set of geo-referenced samples that give the elevation of the terrain. In several applications, the dataset is usually huge, thus yielding high computational times and storage costs. For this reason, it can be useful to build an approximated model in which the precision of the representation is traded for simplicity and speed. The idea consists of selecting a subset of the input dataset and generating a surface representation on the basis of this 'reduced' information. Different degrees of resolution in different parts of the domain are often required in a complex application and the rigidity of an approximated model at a fixed precision can be overcome by developing *multiresolution models*. Research efforts have been devoted to the representation of topographic surfaces in the context of GIS (Gomez and Guzman, 1979; von Herzen and

Barr, 1987; De Floriani, 1989; Samet, 1990a; De Floriani and Puppo, 1992; Samet and Sivan, 1992; Scarlatos and Pavlidis, 1992).

Here we provide a unifying framework for multiresolution representation of terrain surfaces. Relevant examples of multiresolution models (both hierarchical and pyramidal) proposed in the literature are formalized and discussed in the context of such a framework. We will also mention generalizations of the multiresolution approaches for representing hypersurfaces defined by multivariate scalar fields at different resolutions. The remainder of this chapter is organized as follows. In section 1.2, the concept of digital terrain model is formalized. In section 1.3, the problem of terrain approximation is considered and related open problems are outlined. In section 1.4, a formal definition of multiresolution models is introduced that allows a systematic description of all proposed structures. In section 1.5, the state of the art in the context of multiresolution topographic surface description is illustrated and further extensions of this work are proposed. In section 1.6, some concluding remarks are outlined and future developments are presented.

1.2 TERRAIN MODELS

Given a domain $\Omega \subseteq R^2$, and a function $f: \Omega \to R$, the surface corresponding to f over W is mathematically described by the graph of f, i.e., by the set $\{(x, y, f(x, y))/(x, y) \in \Omega\}$. We call the pair $\mathcal{M} = (\Omega, f)$ a *mathematical terrain model*. In practical applications, function f will be *a priori* unknown, while its value will be obtained by sampling the terrain at a finite set P of points in the domain Ω. Thus, input consists of a finite set of *scattered positional data* $\{p1, .., pN/\forall i = 1, .., Np_i = (x_i, y_i, f(x_i, y_i))$ with $(x_i, y_i) \in P\}$ plus possibly additional directional data corresponding to derivatives of f at points of P.

There exists a wide literature concerning the reconstruction of topographic surfaces from a given dataset P. Two basic approaches have been proposed (Foley and Nielson, 1994): the *distance-weighted approach* and the *domain-subdivision approach*. The distance-weighted approach is based on the definition of a single analytic function F interpolating f at the given set P. In this work, we follow the domain-subdivision approach that is based on the definition of a piecewise polynomial function F interpolating, or approximating, f at P over a subdivision of domain Ω into polygonal regions.

A subdivision Σ of a set of points V is a plane connected regular straight-line graph having vertices at V, which can be represented by a triple $\Sigma = (V, E, R)$, where the pair (V, E) is the above mentioned graph, and R is the set of plane polygonal regions induced by such graph (Preparata and Shamos, 1985). The portion of plane covered by the regions of R is called the *domain* of Σ and it is denoted by $D(\Sigma)$. The piecewise function $F = \Sigma_{\{\sigma \in R\}} p\sigma\chi\sigma$ represents f on each region σ composing Σ, where p_σ is a function defined over σ, and χ_σ is the characteristic function of σ. A common choice for the p_σ's is to consider k-degree polynomials interpolating function f at the vertices of the corresponding region σ such that the resulting function F is a piecewise k-degree polynomial. As a consequence, since f is known only at P, the set V of vertices of subdivision Σ is selected to be the same of P.

Let $\mathcal{F} = \{p_\sigma/\sigma$ is a region of $\Sigma\}$. The pair $\mathcal{D} = (\Sigma, \mathcal{F})$ is a *digital model* of f (also called a *digital terrain model*).

The topology of the subdivision underlying a digital terrain model can be either regular or irregular, hence classifying such models into two major classes: *regular* and *irregular* models. The regularity of the subdivision (usually into squares or rectangles (Samet, 1990a,b)) is a great advantage from a computational point of view. Also, early research on surface interpolation was done on regularly distributed data (see Farin (1988) and references therein). A major drawback of regular models is the requirement on data distribution that restricts their applicability to gridded data. Irregular models, on the contrary, do not impose any requirement on both data distribution and domain shape. Triangulations are commonly used as underlying subdivisions for irregular models (see Lee (1991) for a survey).

The properties of functions p_σ composing family $\mathscr{F}$ and their behaviour on edges common to adjacent regions of Σ have a great impact on the shape of the resulting function F. Several approaches have been proposed in the literature for selecting a space of functions for the p_σ's with the aim of defining a piecewise polynomial function F defined over a triangulation Σ with polynomial degree k and order r of smoothness. *Local* approaches defined each p_σ on the basis of the positional and directional information at the three vertices of each triangle σ of Σ (Akima, 1978). A drawback of local methods is the high degree of the resulting piecewise function. *Global* methods have been proposed with the shortcoming that the construction of function F requires the solution of a large (even if sparse) system of linear equations (Farin, 1983; Gmelig *et al.*, 1990).

1.3 APPROXIMATED TERRAIN MODELS

The idea underlying approximated models is that of generating a representation of f on the basis of a selected subset of dataset P. The concept of *precision* in the approximation involves a comparison between the characteristics of the approximated model and the available information about f. A common approach in defining the precision in the approximation consists of considering the maximum distance between F and f at the points of P. In what follows, given a digital model $\mathscr{D}$, the error in approximating f through $\mathscr{D}$ will be denoted by $E(\mathscr{D})$. Also, for approximated models, a classification can be made on the basis of the topology of the underlying domain partition Σ. Irregular models are clearly better suited as a basis for an approximated model, since they do not require any specific data distribution and point selection can be carried on according to an error-driven criterion. In the following, $\mathscr{D}$ is assumed to be an irregular model defined on a triangulation Σ.

The problem of approximating a scalar field f by means of an approximated model $\mathscr{D}$ can be formally stated as follows: given a real value $\varepsilon > 0$, find a digital model $\mathscr{D} = (\Sigma, \mathscr{F})$ as simple as possible, such that $E(\mathscr{D}) \le \varepsilon$. The concept of simplicity relies on the characteristics of both Σ and $\mathscr{F}$ with respect to the amount of information that is actually used: for a given approximation error, it is desirable to use the triangulation and the space of functions for the p_σ's defining a digital model with the smallest complexity (i.e. with the minimum number of vertices in Σ) and providing the best surface shape (i.e. the highest order of smoothness, once the degree of polynomials in $\mathscr{F}$ is fixed). Thus, the following degrees of freedom have to be considered whenever approximating f with $\mathscr{D}$: the triangulation Σ, and the interpolant p_σ to be used on each triangle σ of Σ. In the literature, such approximation problem has usually been faced by fixing the space of functions for the p_σ's (linear interpolants are commonly adopted), and by interacting only with the selection of the triangulation. Two further degrees of freedom characterize a triangulation Σ: its vertex set V, and its connecting structure. A common approach consists of fixing one such degree and optimizing with respect to the other.

Given a vertex set V, there are exponentially many triangulations having vertices at V and it is not known how to compute the triangulation providing the minimum error without considering the whole search space. Criteria have been defined that minimize other quantities, characterizing either the geometry of the triangulation or the shape of the resulting surface instead of the error of the resulting approximated surface. Among geometric criteria, we recall the *Delaunay* criterion that maximizes the minimum angle of its triangles (Preparata and Shamos, 1985). *Data-dependent* criteria (see Dyn *et al.*, 1990: Rippa, 1990, 1992; Quak and Schumaker, 1990; Schumaker, 1993) depend on the z-values of points in V instead of simply on their $x-y$ coordinates, and express certain properties of the resulting surface.

If we fix the connecting structure of the triangulation Σ, the problem of selecting a vertex set $V \subseteq P$ as small as possible such that $E(\mathcal{D}) \leq \varepsilon$ has been investigated in the literature. Two major heuristic approaches have been proposed, corresponding either to a refinement, or to a simplification technique. The refinement approach starts with an initial triangulation of a small set of vertices and refines it by adding new vertices until the given precision is achieved (Fowler and Little, 1979). The simplification approach is based on a decimation technique that, starting from a triangulation involving the whole set P, iteratively eliminates as many vertices as possible along with the edges incident in such vertices, as long as the required precision is satisfied (Lee, 1991; Schroder *et al.*, 1992; Turk, 1992; Hoppe *et al.*, 1993).

1.4   MULTIRESOLUTION TERRAIN MODELS: BASIC DEFINITIONS

In this section, we formally define the concept of multiresolution terrain model that captures all the most important examples of multiresolution models proposed in the literature for describing topographic surfaces at different levels of detail. A multiresolution terrain model is composed of a sequence of surface representations at increasingly finer levels of detail. We focus on the structure of domain subdivision by introducing the concept of multilevel subdivision.

A multilevel subdivision is based on a finite sequence of subdivisions having vertices at a set V of points in R^2: each subdivision can be seen as a refinement of the subdivision preceding it (or, equivalently, as a simplification of the subdivision following it in the sequence). Formally, let D be a compact domain of R^2, and let V be a finite set of points in D. A *multilevel subdivision* having vertices at V is a collection $\mathcal{S} = \{\Sigma_0, \ldots, \Sigma_h\}$ of subdivisions such that $\forall j = 0, \ldots, h$, $\Sigma_j = (V_j, E_j, R_j)$, $D(\Sigma_j) \equiv D$, $V_j \subseteq V$ and $\forall k = 1, \ldots, h$, $V_{k-1} \subseteq V_k$. Figure 1.1 shows an example of a multilevel triangulation.

A region σ may belong to different subdivisions in sequence $\mathcal{S}$. If Σ_i is the first subdivision at which σ appears, i is said to be the *creation level* of σ. The *threshold value* of σ corresponds to the first level greater than i at which σ disappears.

Pairs of regions belonging to consecutive subdivisions in sequence $\mathcal{S}$ may have a spatial interference, i.e., a region σ_k of Σ_i may interest a region σ_l *of* Σ_{i+1}. A special case of multilevel subdivision consists of a sequence $\mathcal{S}$ of subdivisions in which interference relations between pairs of regions belonging to consecutive subdivisions

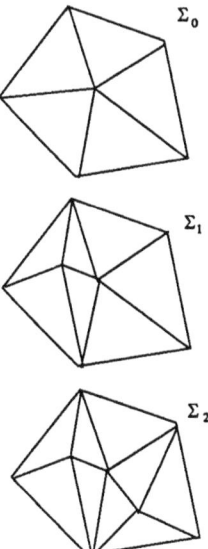

Figure 1.1　An example of a multilevel triangulation.

reduce to inclusion relations. Such multilevel subdivision, that we will call a *hierarchical subdivision*, corresponds to the situation in which the refinement criterion is locally applied to each single region instead of globally to the whole domain. Figure 1.2 shows an example of a hierarchical triangulation.

The concepts of terrain modelling (defined in sections 1.2 and 1.3) and of multilevel subdivision (above described) can be combined in order to define a multiresolution terrain model. A multiresolution terrain model is simply a terrain model based on a multilevel subdivision. Usually, a decreasing sequence of tolerance values

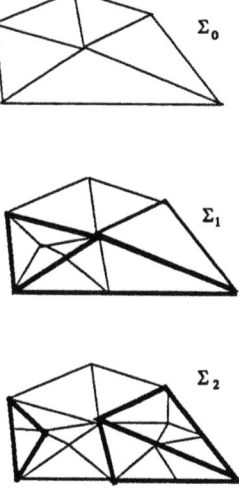

Figure 1.2　An example of a hierarchical triangulation.

$[\varepsilon_0, .., \varepsilon_h]$ corresponding to progressively finer levels of detail is provided. A multi-resolution model is defined in such a way that each Σ_i approximates the surface at precision ε_i (with $0 \le i \le h$).

1.5   MULTIRESOLUTION TERRAIN MODELS: STATE OF THE ART

A few proposals of models for multiresolution description of topographic surfaces can be found in the literature. All proposed models are essentially hierarchical structures that guarantee multiresolution by interacting on domain partition, while *a priori* fixing the space of functions. Little work has been done for general pyramidal models. Multiresolution models can be classified according to domain partition, to spatial relations between different representations, to the possibility of guaranteeing a continuous description at each level, to the satisfaction of certain geometric properties, etc. A formal definition and an analysis of multiresolution terrain models as well as a description of algorithms for building them can be found in (De Floriani *et al.*, 1996).

The *quadtree* (Samet, 1990a) is a well-known hierarchical data structure for two-dimensional spatial data, in which an initial square domain is partitioned into a set of nested squares by recursive split of each square into four quadrants. A quadtree-based surface model can be obtained by selecting vertices of squares at data points, and by using bilinear interpolation to approximate the surface inside each square. In this case, the rule for building the quadtree is the following: any square exceeding a given tolerance value ε is split into four subsquares by joining its centre to the midpoints of its four sides (Figure 1.3a). Chen and Tobler evaluate alternative interpolation techniques for approximating a surface defined by a quadtree in terms of accuracy and efficiency (Chen and Tobler, 1986).

In a *quaternary triangulation* (Gomez and Guzman, 1979; Barrera and Vazquez, 1984), the initial domain is a triangle containing all data points. At each level of recursion, a triangle, that does not satisfy the tolerance value ε, is subdivided into four subtriangles formed by joining the three midpoints of its triangle sides (see Figure 1.3b). The terrain surface is approximated inside each triangular patch by linear interpolation at the elevation values of the three vertices.

Both quadtrees and quaternary triangulations require regularly spaced data points. Besides the need for regularly distributed data, a quadtree or a quaternary triangulation could not preserve the continuity of the approximation along the edges of the subdivision. Figure 1.4 shows an example of discontinuity in a

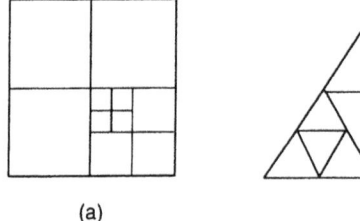

(a)   (b)

Figure 1.3   An example of quadtree (a) and of quaternary triangulation (b).

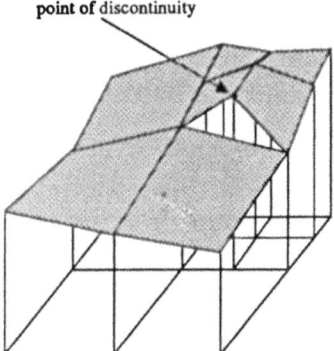

Figure 1.4 An example of discontinuity in a surface quadtree.

quadtree. Several techniques have been suggested in the literature for coping with the problem of discontinuities in quadtree-based models (Barrera and Hinojosa, 1987; von Herzen and Barr, 1987). The most popular method is the *restricted quadtree* (von Herzen and Barr, 1987; Samet and Sivan, 1992).

The first example of hierarchical model based on triangles is the *ternary triangulation* (De Floriani *et al.*, 1984; Ponce and Faugeras, 1987). Such representation is suitable for irregularly distributed data, and it is based on the recursive subdivision of an initial triangle, covering the whole domain, into a set of nested subtriangles with vertices at the data points. A subdivision of a triangle t is obtained by joining an internal point p to the three vertices of t (Figure 1.5): point p is the point that maximizes the approximation error inside t. The major problem with a ternary triangulation is the elongated shape of its triangles, which leads to inaccuracies in numerical interpolation. Moreover, in a ternary triangulation, like in quadtree-based models (i.e. in quadtrees, quaternary triangulations and restricted quadtrees), an explicit representation at different levels of precision is not supported.

Recently, hierarchical triangle-based models that explicitly encode a sequence of descriptions at different resolutions have been defined. Usually, a sequence of tolerance values corresponding to such levels is specified. The *adaptive hierarchical triangulation* (Scarlatos and Pavlidis, 1992) has been proposed with the aim of retaining cartographic coherence in the hierarchical representation. The tree is constructed in such a way that the root provides a surface approximation at the coarsest level of precision. At any further step, each triangle, which does not satisfy the current tolerance, is recursively refined into a triangulation by inserting one vertex at a time, until this precision is met. Inserted points are a subset of data corresponding to surface specific points and are extracted from the dataset according to a heuristic *splitting rule* that permits to split also edges of triangles. In the adaptive hierarchical triangulation, the problem of 'sliveriness' is not completely cured, since

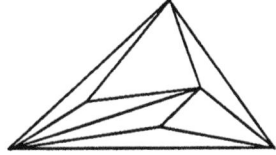

Figure 1.5 An example of a ternary triangulation.

the triangulation pattern used for driving refinement does not take into account the shape of triangles explicitly.

The approach proposed in (De Floriani and Puppo, 1992), called the *hierarchical Delaunay triangulation*, tries to solve such problems by making local use of the Delaunay triangulation. The use of Delaunay triangulation not only helps to maintain the shape of triangles as closely equiangular as possible, but also provides an approximation with other interesting geometric properties: as shown (in Rippa, 1990), Delaunay triangulation minimizes the roughness of the approximating surface. The refinement scheme for generating the hierarchical structure is the same that characterizes the adaptive hierarchical triangulation, while a refinement technique for point selection and update of a Delaunay triangulation is used, instead of the splitting rule. Adaptive hierarchical triangulations and hierarchical Delaunay triangulations can be generated either from regular or irregular data distributions. Moreover, a continuous representation at each level of precision in the given sequence of tolerance values can be easily extracted.

Little work has been done in the literature in the context of pyramidal models. The first proposal in this area is represented by the *Delaunay pyramid* (De Floriani, 1989), which is based on a sequence of Delaunay triangulations satisfying progressively finer precision levels. An efficient data structure for pyramidal models has been proposed, which is called the *sequence of lists of triangles* (SLT) (Bertolotto *et al.*, 1994a): such structure stores minimal topological information about entities belonging to the same level, while it encodes links between entities belonging to different levels in the model to describe their spatial interference. More recently a data structure has been proposed that stores a pyramidal triangle-based terrain model by avoiding duplications, while maintaining topological information between entities at the same level, and without encoding interference links (Cignoni *et al.*, 1995). Such a data structure is especially well suited for visualization at variable resolution over the domain. An alternative pyramidal model based on Delaunay triangulations has been proposed that is built according to a simplification technique (de Berg and Dobrindt, 1995). Such a model provides an efficient structure for solving geometric queries (such as point location) and supplies visualization at a variable precision on different parts of the domain. Unlike the Delaunay pyramid, it does not guarantee explicit multiresolution at increasingly finer levels of detail.

Some models described for terrains have been extended for multiresolution representation of three-dimensional (and, more recently, multivariate) scalar fields. Three-dimensional generalizations of surface quadtrees can be obtained on the basis of *octrees* (Chien and Aggarwal, 1986; Samet, 1990b), which require regularly distributed data and provide a domain subdivision into regular three-dimensional cells (cuboids) of variable dimensions. A hierarchical model for the representation of three-dimensional scalar fields based on tetrahedralizations has been proposed in (Bertolotto *et al.*, 1994d). This model is an extension of the hierarchical model for representing two-dimensional scalar fields proposed in (De Floriani and Puppo, 1992): the use of irregular cells allows the possibility of adapting to hypersurface irregularities and modelling scattered volume data. A first attempt to multiresolution description in the generic d-dimensional case has been recently done in (Bertolotto *et al.*, 1994b) by extending the model to d-variate scalar fields. In (Bertolotto *et al.*, 1995), a pyramidal model for representing hypersurfaces defined by multivariate scalar fields has been proposed, which directly extends the presented definition of pyramidal terrain model.

1.6 FUTURE PERSPECTIVES

A related issue of great impact in GIS is the design of models for representing geographic maps at multiple resolution. Concepts developed in designing multiresolution terrain models help in investigating multiresolution for maps, though more general problems can arise. Spatial subdivisions in terrain modelling are just a means of discretizing the domain in order to obtain locally simple representations of the surface, while such discretization has in general no relevance in defining the structure of the unique entity represented, i.e. the terrain. On the contrary, geographic maps are intrinsically characterized as objects composed of simpler entities; hence, a subdivision underlying the representation of a map must reflect the structure of the map itself. For this reason, quite general tessellations formed with regions of arbitrary shape, and possible point and lineal features must be adopted. Another key point in multiple resolution for maps is relating representations of an entity at different scales; in particular, the shape and dimension of a geometric object representing a given entity may be very dependant on scales (Frank and Timpf, 1994).

A first formal framework for multiresolution maps was proposed in (Bertolotto *et al.*, 1994c); the model is based on a hierarchy of topological representations which extends the concept of hierarchical triangulation. Apart from being based on more general subdivisions, the main difference from hierarchical triangulations is relaxation of the geometric constraint on coincidence of domains between a region and its refinement. In a more recent work (Puppo and Dettori, 1995), the issues involved in the representation of spatial maps at different resolutions have been elaborated more thoroughly, and the possibility has been introduced of representing a geographic entity through geometric objects of different dimension (regions, lines or points), depending on the level of resolution.

REFERENCES

AKIMA, H. (1978) A method of bivariate interpolation and smooth surface fitting for irregularly distributed data points. *ACM Transac. Math. Software* **4**, 148–159.

BARRERA, R. and VAZQUEZ, A. M. (1984) A hierarchical method for representing relief. Proceedings of the Pecora IX Symposium on Spatial Information Technologies for Remote Sensing Today and Tomorrow, Sioux Falls, South Dakota, 87–92.

BARRERA, R. and HINOJOSA, M. (1987) Compression method for terrain relief. Technical Report, CINVESTAV, Engineering Projects, Department of Electrical Engineering, Polytechnic University of Mexico, Mexico City.

BERTOLOTTO, M., DE FLORIANI, L. and MARZANO, P. (1994a) An efficient representation for pyramidal terrain models. Proceedings of the 2nd ACM Workshop on Advances in GISs, Gaithersburg, Maryland, 129–136.

BERTOLOTTO, M., DE FLORIANI, L. and PUPPO, E. (1994b) Hierarchical hypersurface modelling. IGIS'94: Geographic Information Systems, LNCS 884, NIEVERGELT, J., ROOS, T., SCHEK, H. J., WIDMAYER, P., eds, Springer Verlag, 88–97.

BERTOLOTTO, M., DE FLORIANI, L. and PUPPO, E. (1994c) Multiresolution topological maps. Advanced Geographic Data Modelling – Spatial Data Modelling and Query Languages for 2D and 3D Applications, Netherlands Geodetic Commission, **40**, 179–190.

BERTOLOTTO, M., DE FLORIANI, L. and MARZANO, P. (1995) Pyramidal simplicial complexes. Proceedings of the 3rd Symposium on Solid Modelling and Applications 1995, Salt Lake City, Utah, 152–163.

BERTOLOTTO, M., BRUZZONE, E., DE FLORIANI, L. and PUPPO, E. (1994d) Multi-resolution representation of volume data through hierarchical simplicial complexes. Proceedings of the 2nd International Workshop On Visual Form, Capri, Italy.

BRUEGGER, B. and FRANK, A. (1989) Hierarchies over topological data structures. Proceedings of the ASPRS-ACSM Annual Convention, Baltimore, MD, 137–145.

BRUEGGER, B. and KUHN, W. (1991) Multiple topological representations. Technical Report 91–17, National Centre for Geographic Information and Analysis, Santa Barbara, CA.

CHEN, Z. T. and TOBLER, W. R. (1986) Quadtree representation of digital terrain. Proceedings of the Autocarto, London, 475–484.

CHIEN, C. H. and AGGARWAL, J. K. (1986) Volume/surface octrees for the representation of three-dimensional objects. *Computer Vision, Graph. Image Process.* **36**, 100–113.

CIGNONI, P., PUPPO, E. and SCOPIGNO, R. (1995) Representation and visualization of terrain surfaces at variable resolution. International Symposium on Scientific Visualization, Cagliari, Italy.

DE BERG, M. and DOBRINDT, K. (1995) On levels of detail in terrains. Proceedings of the 11th Annual ACM Symposium on Computational Geometry, Vancouver, BC, Canada.

DE FLORIANI, L. (1989) A pyramidal data structure for triangle-based surface description. IEEE Computer Graphics and Applications, **9**, 67–78.

DE FLORIANI, L. and PUPPO, E. (1992) A hierarchical triangle-based model for terrain description. Theories and Methods of Spatio-Temporal Reasoning in Geographic Space, FRANK, A. U., CAMPARI, I., FORMETINI, U., eds, LNCS N.639, Springer-Verlag, 236–251.

DE FLORIANI, L., MARZANO, P. and PUPPO, E. (1996) *Multiresolution Models for Topographic Terrain Description, The Visual Computer* (in press).

DE FLORIANI, L., FALCIDIENO, B., PIENOVI, C. and NAGY, G. (1984) A hierarchical data structure for surface approximation. *Computers Graph.* **8**, 475–484.

DYN, N., LEVIN, D. and RIPPA, S. (1990) Data dependent triangulations for piecewise linear interpolation. *IMA J. Numerical Analysis* **10**, 137–154.

FARIN, G. (1983) Smooth interpolation to scattered 3D data. In *Surfaces in CAGD*, BARNHILL, R. E. and BOHEM, W., eds, North-Holland Publishing Company.

FARIN, G. (1988) *Curves and Surfaces for Computer Aided Geometric Design*, RHEINBOLDT and SIEWIOREK, eds, Academic press, San Diego.

FOLEY, T. A. and NIELSON, G. M. (1994) Modelling of Scattered Multivariate Data. Eurographics '94: state of the art reports, Oslo, Norway, 38–59.

FOWLER, R. J. and LITTLE, J. J. (1979) Automatic extraction of irregular network digital terrain models. *Computer Graph.* 199–207.

FRANK, A. U. and TIMPF, S. (1994) Multiple representations for cartographic objects in a multi-scale tree – an intelligent graphical zoom. *Computer Graph.* **18**, 823–829.

GMELIG MEYLING, R. H. J. and PLUGER, P. R. (1990) Smooth interpolation to scattered data by bivariate piecewise polynomials of odd degree. *Computer Aided Geom. Design* **7**, 439–458.

GOMEZ, D. and GUZMAN, A. (1979) Digital model for three-dimensional surface representation. *Geo-Processing* **1**, 53–70.

HOPPE, H., DEROSE, T., DUCHAMP, T., MCDONALD, J. and STUETZLE, W. (1993) Mesh optimization. Computer Graphics Proceedings, Annual Conference Series, 19–26.

LEE, J. (1991) Comparison of existing methods for building triangular irregular network models of terrain from grid digital elevation models. *Internat. J. Geogr. Inform. Syst.* **5**, 267–285.

PONCE, J. and FAUGERAS, O. (1987) An object centered hierarchical representation for 3D objects: the prism tree. *Computer Vision, Graph, Image Process.* **40**, 1–29.

PREPARATA, F. P and SHAMOS, M. I. (1985) *Computational Geometry: an Introduction.* Springer-Verlag.

PUPPO, E. and DETTORI, G. (1995) Towards a formal model for multiresolution spatial maps. Proceedings of the 4th International Symposium on Large Spatial Databases, Portland, Maine.

QUAK, E. and SCHUMAKER, L. L. (1990) Cubic spline fitting using data dependent triangulations. *Computer Aided Geom. Design* **7**, 293–301.

RIPPA, S. (1990) Minimal roughness property of the Delaunay triangulation. *Computer Aided Geom. Design* **7**, 489–497.

RIPPA, S. (1992) Adaptive approximation by piecewise linear polynomials on triangulations of subsets of scattered data. *SIAM J. Sci. Stat. Comput.* **13**, 1123–1141.

SAMET, H. (1990a) *Applications of Spatial Data Structures*. Reading, MA: Addison Wesley.

SAMET, H. (1990b) *The Design and Analysis of Spatial Data Structures*. Reading, MA: Addison Wesley.

SAMET, H. and SIVAN, R. (1992) Algorithms for constructing quadtree surface maps. Proceedings of the 5th International Symposium on Spatial Data Handling, Charleston, 361–370.

SCARLATOS, L. L. and PAVLIDIS, T. (1992) Hierarchical triangulation using cartographic coherence. *CVGIP: Graph. Models Image Process.* **54**, 147–161.

SCHRODER, W. J., ZARGE, J. A. and LORENSEN, W. E. (1992) Decimation of triangle meshes. *Computer Graph.* **26**, 65–70.

SCHUMAKER, L. L. (1993) Triangulations in CAGD. *IEEE Computer Graph. Applic.* **13**, 47–52.

TURK, G. (1992) Re-tiling polygonal surfaces. *Computer Graph.* **26**, 55–64.

VON HERZEN, B. and BARR, A. H. (1987) Accurate triangulations of deformed, intersecting surfaces. *Computer Graph.* **21**, 103–110.

CHAPTER TWO

A generic large-scale spatial data model for surveyors: changing practices and processes

DAVID PARKER, GEORGE TAYLOR and CRISPIN HOULT

2.1 INTRODUCTION

This chapter discusses a generic data model for large-scale administrative and engineering spatial data. The model requires the standardisation of a basic spatial unit: the land face. The model is designed to enable spatial datasets created now to be increased in resolution and dimensionality as information needs develop.

We are addressing the problems faced by surveyors in providing clients with spatial data for computerised information systems. This is not a once-and-for-all operation. A client usually begins with a simple spatial representation of their features of interest: a point, say, to represent a building. As their needs develop they wish for an improved resolution and the surveyor is asked to provide it. Consider a local authority initially representing a property by a single point but later wishing to represent it by a boundary and possibly with an associated surface model. What must be avoided is the situation where the dataset has to be rebuilt every time increased resolution or update is required. The ideal is for variable resolutions of spatial data to exist in an information dataset at one time with the possibility to increase the spatial resolution of any feature upon demand.

To put the work in context, it is worth keeping in mind the range of applications for spatial information datasets within a local authority. They record a range of information against a background of Ordnance Survey (OS) 1:1250 sheets: land holdings, landscaped areas needing maintenance, road areas needing surfacing, locations of street lighting and road signs. They control planning applications affecting properties in their area. They design new road schemes and municipal building requiring larger scale spatial data (e.g. 1:250) including ground relief. They have internal plans of all the floors of the buildings they maintain. Note that most of their applications consider area features.

2.2 LARGE-SCALE MAP DATA FOR INCORPORATION IN MULTI-FUNCTION INFORMATION SYSTEMS

To most map users in the UK the term *large-scale* refers predominantly to data at the basic map scales of 1 : 1250 and 1 : 2500 from the OS. The OS is currently the dominant source of such data but this may not always be the case. These OS datasets were designed largely for administrative and cartographic purposes and have many shortcomings when used for applications such as asset recording, planning, engineering and landscaping. Many applications such as these require a higher resolution (larger scale), greater dimensionality (additional 'floors', a ground surface model or full 3D) and many additional attributes.

Incorporation referred to above means more than utilising map data as a back-drop: a spatially precise wallpaper on which to hang other information such as utility plant records. Incorporation means the annexing of the geometrical elements, their attributes and previously constructed topology, possibly in association with new geometry and attributes, to create 'objects' of interest to a user. It is important to realise that the great majority of objects needed by users of large scale applications are area objects: buildings, plots, roads, fields, car parks. The complexity in incorporation can be great. A user creates objects utilising a surveyor's raw data set. The objects, perhaps land parcels or segments of road, may be further annexed by other users to create objects of higher intelligence, say an area of housing or a complete transport route. Establishing formalised methodology for the capture and modelling of spatial data to achieve all perceived user requirements is a complex and difficult task.

The surveyor supplies spatial data to all no matter what recording and analysis software they use. Although, at present it is not an absolute necessity for individual in-house stand-alone information systems to have the ability to cross reference one user's data set with another, it will soon become essential. The data models used by a surveyor are not the same as those used by individual clients. Nevertheless, a surveyor's data model must be consistent with the model requirements of each individual client.

2.3 WHAT THE SURVEYOR MUST SUPPLY TO THE CLIENT

It would be most beneficial if a core spatial data set existed on which all applications can be based. A core data set should minimise data management whilst maximising integration. What digital spatial reference is appropriate? To be realistic it must be one which is available and manageable now but must be able to be refined over time. A progression over time might well start with a single co-ordinated point viewed against a map background. As information needs are refined the progression might well be: a bounded area at say 1 : 1250; a bounded area at larger scale; more than one layer/floor within a bounded area; a surface model within the bounded area; a 3D model with plan consistent with the bounded area.

What is the current position? With the availability of such datasets as ADDRESS-POINT, many organisations are embarking on the first stage in this continuum: the use of a single co-ordinated point viewed against a map background. However, some disciplines within an organisation are already at a stage which requires the use of bounded areas. An ideal is to ensure that the data on which they build their applications can be utilised by others as their information needs are refined. Another requirement is to ensure that the data set can be sensibly maintained as change occurs. This change can be as a result of object modification but it can also be as higher resolution or dimensionality of the data set is required. This is most likely to occur in limited localities at one time. The surveyor's practices and generic model must cope with this.

2.3.1 Two examples

Two examples will serve to demonstrate what the client requires now and how this requirement will change in the future as information demands develop. Some terms such as 'land face' and 'dark links' are introduced here and then defined in more detail later.

Consider a block of housing as represented on a large scale administrative map, Figure 2.1a. The surveyor's representation proposed below is that of a series of land faces, one for each division of building, garden, yard and pavement. A property terrier application is concerned only with ownership, consequently complex objects of land parcels would be created. A land parcel in this example would be the contiguous area defined by three faces: the front garden, the building and the back yard, Figure 2.1b. A land use application is concerned only with the contiguous area of residential parcels, Figure 2.1c. In order for the land use application to maintain a precise link with the terrier application and the surveyor's data, there are three options: build a complex object from the surveyor's land faces; build complex objects from the parcels of the terrier application; or build a container object, a

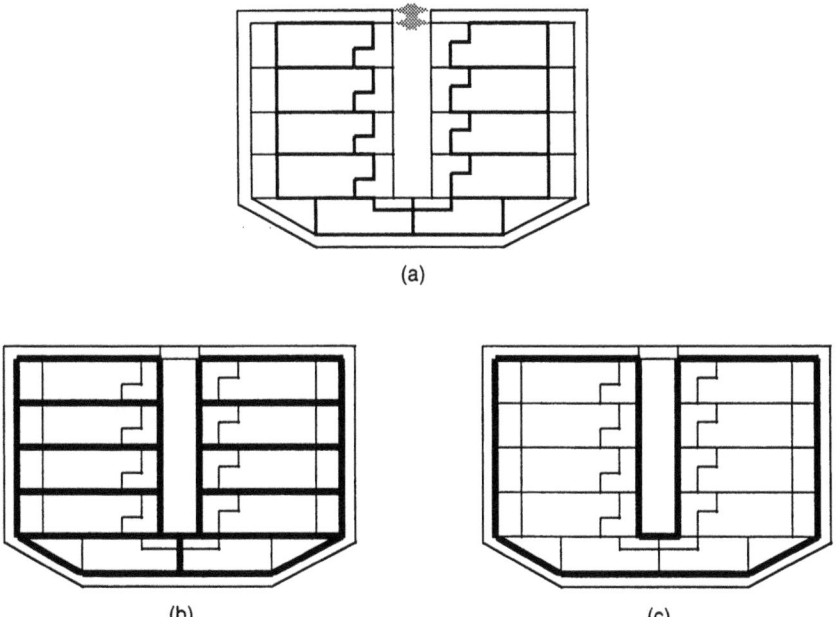

(a)

(b)   (c)

Figure 2.1   (a) Block of housing; (b) contiguous areas defined by three faces; (c) contiguous area of residential parcels.

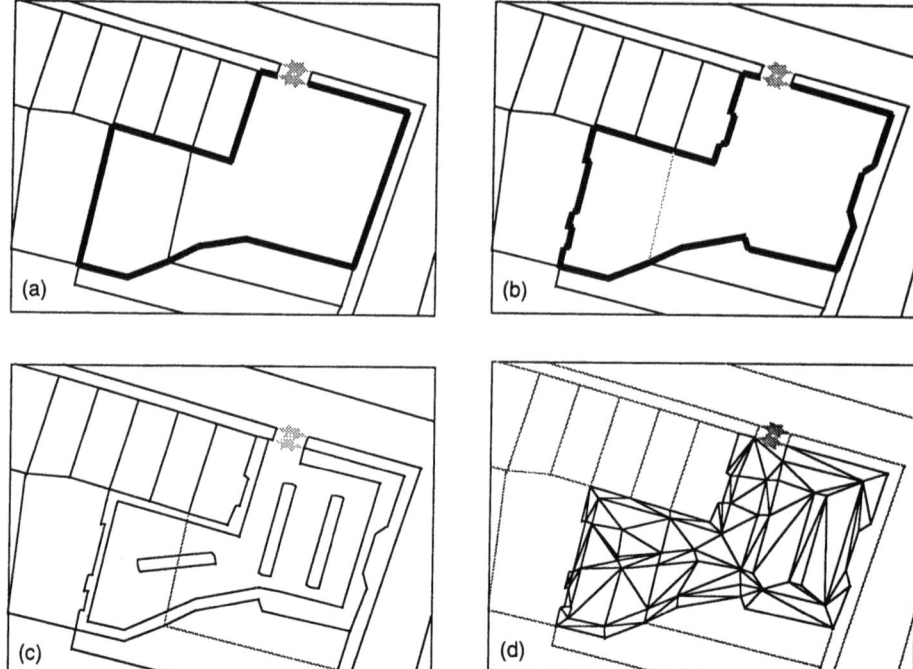

Figure 2.2 (a) A series of land faces; (b) faces defined to a higher resolution; (c) further geometrical plan details; (d) surface model within a face.

boundary only, from either the surveyor's faces or the terrier parcels (the resultant defined area would be the same).

The second example addresses the update problem. Figure 2.2a represents a series of land faces in an urban area. The two faces in the centre have been incorporated within a single parcel in the property terrier. The parcel has been redeveloped into a car park and a record of the layout, landscaping and drainage is required. The information is available as an 'as-built' plan at 1 : 250 scale with an associated triangular irregular network (TIN) model. This new information must be incorporated into the information database. What changes are necessary in the spatial representation of the terrier? The parcel is conceptually the same but the boundaries of the faces from which it has been built are now defined to a higher resolution, Figure 2.2b. An adaptable data model must be able to refer back to the surveyed elements of the boundary to ensure this increase in resolution is recorded. Further geometrical plan details are now available within each face and should be associated with that face, Figure 2.2c. Similarly, a surface model is now available within each face and again should be associated with that face, Figure 2.2d.

2.4 SURVEYOR'S CURRENT PRACTICES

The surveyor's traditional practice has been to sample the 2D position of visible boundaries and other linear features on the ground (natural and manmade), coding these and presenting the result as a graphic. In association with this would often be surface relief information represented on the graphic as contours. More recently, the graphic would be digital and the relief information modelled, for example as a TIN, though generally independent of the plan. In the process of survey for engineering and similar works, the measurements for the planimetric detail and the surface model are done together: in fact, the sample points selected to represent the locations of planimetric detail and to build the model of surface relief are usually the same.

2.5 DEFINITION OF A LAND FACE

Plan information as a digital graphic is still a considerable way from the raw spatial data required for incorporation in an information system: a data set guaranteed able to be constructed into user definable objects. A contiguous surface cover of land faces has been proposed as the appropriate model for this (Bridger and Land, 1992), (Hoult and Parker, 1993), (Parker *et al.*, 1994). The AGI GIS Dictionary (AGI, 1994) defines a face as 'A surface bounded by a closed sequence of edges. Faces are contiguous and fill the spatial extent of the dataset and do not overlap'. The basic cover of land faces would have to be accompanied with other 'layers' of such as utility information, individual items of vegetation etc. The idea of land faces was introduced in the examples earlier. We now look at the requirements for a data model built from such faces.

2.6 REQUIREMENTS FOR A DATA MODEL

What should be the design requirements of a data model based on land faces, which is practical for surveyors to implement and maintain? Most of the requirements were demonstrated by the two examples described earlier, and it is useful here to refer back to Figures 2.1 and 2.2.

- What should constitute the basic face unit and how should they be classified? This classification must be consistent with OS large scale data. The classifications used by the OS for Project 93 were building, vegetation, landform, water, road and miscellaneous (Ordnance Survey, 1993). The first three of these are already implied in OS Land-Line data. Provided clearly defined standards can be agreed, assigning a classification to a face is not a problem. Developing a standard usually is.
- Where should dark links be positioned? Dark links represent the many boundaries in the real world where a road object changes to a car park, a school playing field or a private drive etc., which can be interpreted by eye, but which are not explicitly defined in the surveyed data (Bridger and Land, 1992). Provided the recommended positions are clearly defined by a standard, surveyors can work to it. Again, a standard is required.
- Dark faces must be permitted: a dark face is where one face overlaps another face. An example might be where one road bridges another. The recording of such as multi-level shopping complexes also requires dark faces.
- It must be possible to increase the resolution of face boundaries without effecting the definition of the objects built from the faces.

- When an increased 'scale' of resolution is required it is normally required to associate additional 'internal' geometry with a face.
- When the resolution is increasing for engineering works it will be necessary to associate a TIN model with a face. The boundary of the TIN model will be that of the face and the locations of the sample points will be those of the boundary and internal geometry.
- The model must consider multiple 'layers' to a face so, for example, the floors of a building can be referenced to a building face. The boundaries of the secondary layers need not coincide with those of the land face enabling building overhangs or indents to be considered.
- Taking the concept of multiple layers a stage further would require the association of a full three-dimensional model to represent such as a building located on a face.
- Surveyors capture data by field survey, by GPS, by photogrammetric methods from aerial images and by digitising existing documents. Processing practices need to be consistent no matter what the data source.
- Surveyors capture data at a range of 'large scales' – $1:100$ down to $1:2500$. Practices need to be standard across the range of scales.
- A single set of procedures and processing stages is necessary, no matter what the requirements of the recipients information system: whether it be layer-based CAD, a coverage model or an object-based model.
- There is a need to associate metadata and attribute data with both line geometry defining face boundaries and with basic land faces.
- It must be recognised that the coverage of basic land faces would have to be supplemented by other 'layers' of spatial information: utility networks, coverages of points for items of vegetation.

2.7 THE PROPOSED MODEL

The generic face-based spatial data model which considers all these factors is shown in Figure 2.3. The key entity in the model is the land face. The boundary of this is made up of any number of segments of line geometry. Each line geometry entity is associated with two and only two faces. It is possible to associate any number of metadata attributes with both the line geometry and the land faces.

Any number of geometrical entities representing the internal geometry can be associated with a face. If the sample points of both this internal geometry and those of the segments of line geometry describing the boundary of the face are defined three-dimensionally, then associating a 'method' with the face will enable a unique surface model to be constructed.

Any number of secondary face entities can be associated with a single land face. In the case of dark faces, the spatial extent would be that of the land face. More generally, when the secondary face is describing an object such as a building floor, the spatial extent of the entity would be defined by a new closed line geometry. A full 3D model would be built from the land face and a collection of secondary faces.

Container entities are used to define the boundaries of objects of large extent where it is essential that the boundary is coincident with those of faces but association with the faces themselves need not be maintained. A container entity is constructed from any number of face boundary geometry entities.

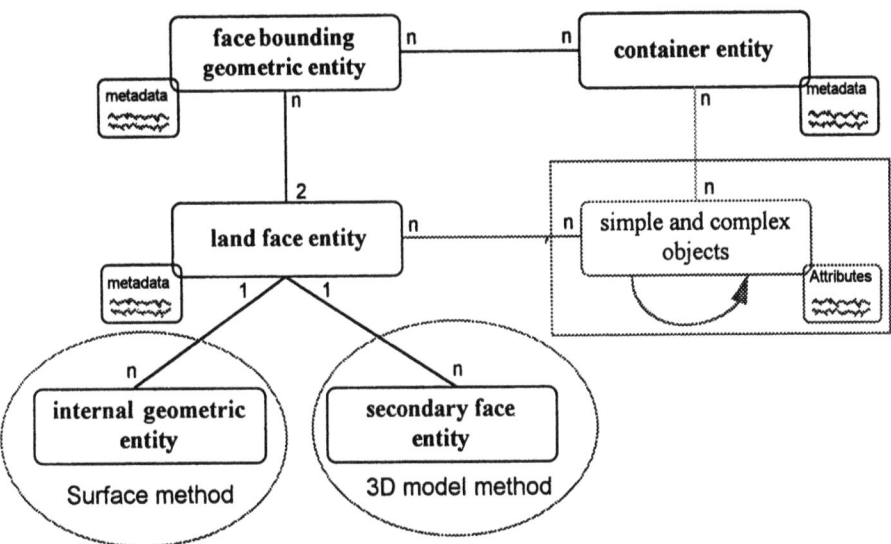

Figure 2.3 A face-based model.

Simple objects are those built from a single face or a single container entity requiring only the association of descriptive and metadata attributes. Complex objects are those built from a collection of faces, container entities or other objects. The spatial extent need not be contiguous and full. Descriptive and metadata attributes are again associated with these objects.

2.8 TRIAL IMPLEMENTATION

The process of capture, management and onward transfer of face-based spatial data to the defined model have been tested at Newcastle. AutoCAD was the choice of software on which to trial the implementation. This software has all the standard CAD features, from symbols and linetypes to advanced 3D modelling entities. Controlled by a graphical user interface, the whole system from menus to commands may be tailored and customised. The 'C' programming language libraries of the application program interface functions allows fast and direct access to drawing creation, editing and selection functions.

AutoCAD is widely used as a graphical design, editing and visualisation engine in surveyor's mapping and GIS solutions. Although AutoCAD has no integral ability to understand topology, this may be managed in AutoCAD by using additional entity tags called extended entity data and the facility for maintaining unique identifiers on entities. This allows complex topology to be managed, e.g. handling of NTF level 3 boundary-line data (AutoNTF, 1994). Holding both graphic and attribute data, the drawing file on disk (DWG) is referred to as the drawing database. Two extensions to AutoCAD were used in this implementation. The AutoCAD Database Extension (ADE) allows a distributed array of drawings to be visualised and queried as a seamless map database. The AutoCAD SQL Extension (ASE) enables SQL compliant databases to be created, linked to drawing entities, populated and searched wholly from within AutoCAD.

To make a face-based dataset:

- Input of 'map' linework. The linework delimiting face boundaries are brought into the CAD facility in any available manner. In Newcastle, all the standard flowlines of field survey, air survey, digitising original paper maps and importing existing digital map data all result in an AutoCAD drawing. This drawing is the raw 'database'. If the 'seeding' of faces was included in the data capture stage, then these should be input to the CAD facility at the same time. Seeding is simply locating a point entity within the boundary of each face coded with the classification of that face.
- Addition of dark links. Dark links are normally required to make it possible to logically close off spatial areas for faces. Dark links are often required to 'close' small gaps such as gates, where missing property fences would rightly be and to segment at junctions of long sections of road.
- Coding of linework (addition of a key descriptive attribute). As will be seen, the faces are to be coded as a separate operation and this alleviates the need for significant coding of the raw linework. Only if a linear feature is to exist in its own right is there a need to code line geometry. Being able to omit the addition of line codes at the capture stage has the potential to greatly increase productivity at this time (but creates some limited additional work later).
- 'Clean'ing and 'Link'ing. The linear geometry of the boundaries and dark links must be cleaned (have undershoots and overshoots removed) and linked (be broken into discrete entities at crossings). For large datasets this is a very time consuming process and preferably should be done in batch mode. Real time cleaning and linking is available however when line geometry is edited.
- Line geometry quality metadata. The addition of quality metadata to the line geometry at the data capture stage and its maintenance throughout processing. This option is yet to be implemented.
- Seeding. The faces must now be seeded in preparation for the building of face topology. Seeding may have occurred at the capture stage but is also easily and quickly achieved in 'head up' mode whilst viewing the line geometry.
- Face building. For each seed, the bounding line geometry are automatically identified. The geometry of the line work is left unaltered but a unique face identifier is added as an attribute to all bounding lines. The face identifier attributes are added as extended entity data. A relational table is built with the unique key of the face identifier and fields of the face classification and quality metadata. Unlike other systems the seeds may now be discarded. Furthermore, due to the use of the AutoCAD graphics engine, other factors traditionally requiring consideration, such as direction and ordering of polygon component links, are also made redundant.
- Face and islands verification. All line geometry are checked to ensure they bound two and only two faces. All faces are checked for overlap. Problems may arise in the face building process; these occur if the line geometry was not correctly cleaned and linked. Islands not found in face making can be easily located at this stage by an automated batch process.
- Update. The topology building and verification are best completed in batch mode but facilities are available to permit easy update of the data in real time.
- Containers. The implementation allows for the creation of containers. To create a container, as in the example of Figure 2.1, the faces abutting the inside of the

container boundary have to be selected. An automatic process then identifies those line entities required to bound the container. These line entities are marked in a similar way to those of land faces and from then on a container is treated in the same fashion as a land face.

- Objects. Creating objects from faces becomes an application dependant operation. With many users of the spatial source of face data it is important to keep the management of this information separated from the user database of specific objects, the geometry is independent. This allows the map maintenance to be limited to the role of the surveyor who may update the spatial element without concern over the wide range of users. Objects are therefore built in the database by a series of relational tables. Complex objects use hierarchy to establish strict parent/child relationships, e.g. a parcel is made up of a land face and a house object.

2.9 EXTRACTING DATA FROM THE DATA STORE

Using the facilities of AutoCAD and ADE, the data may be queried and the resultant subset prepared for export to another system. Queries can be built limiting spatial extent, graphic entity type, any metadata or descriptive attribute (or any combination of these). Generally it is not necessary to maintain topology in the transfer as the data is geometrically clean and topology can be easily rebuilt in the recipient system. If data are required in a CAD layer format then producing a de facto standard DXF data transfer file is straightforward. If data are required to construct coverages in ARC/INFO or objects in SmallWorld then the required proprietary format can be created using the generic output command which allows user defined file structures to be formatted using the current drawing data. If full topology is to be maintained then transfer must be through a standard supporting medium such as BS7567/NTF (BSI, 1992).

2.10 UPDATE

With the supply of spatial data based on land faces, digital map update can easily utilise change-only data. Change-only update is an efficient form of map data management which also allows simple versioning by incorporating time-tag metadata. Face-based change-only data can be easily supplied by the surveyor in a compact and complete form.

The update process works by replacing all linework that is common to all faces that have been updated. Updates operate as deletions and additions: all old faces are removed to a version store and all new information is used to replace these. In simple updates of geometry, where the topology does not change, face id's may remain the same. However, if faces are completely removed from the system (such as the removal of a building) then the database is informed, the old face is given a deleted date and all complex objects referencing that face must have some manual interaction to ensure consistency in the database. Container objects that are affected may be found by spatial queries.

2.11   IS THIS ALL PRACTICAL?

We have discussed a generic data model for large-scale administrative and engineering spatial data. The model requires the standardisation of a basic spatial unit: a land face. The model is designed to enable spatial datasets created now to be increased in resolution and dimensionality as information needs develop. All the principal processes described here have been implemented using a trial data set. Yes, it is all practical and possible.

However, there are two major problems. The standardisation of the land face is one: achieving a standard is never easy. Changing surveyor's working practices is the other: there is inertia in the existing data supply line which must be overcome. If these two problems can be successfully overcome, then it is possible to have datasets of mixed resolution, improving with time and maintained only with the supply of change-only data. For example, a dataset of parcels, certain of these can be represented with a single point, some (where development is planned) by faces at a low spatial resolution and others (where developments are complete) defined by faces with high spatial resolution, with additional internal geometry and a surface model if required. Resolution independent mapping has long been a cartographer's and land surveyor's dream. Could a face-based spatial data model be the answer to this dream, or just another nightmare?

REFERENCES

AGI GIS DICTIONARY (1994) http://www.geo.ed.ac.uk/root/agidict/html/welcome.html

AUTONTF (1994) AutoNTF Manual, Department of Surveying, University of Newcastle upon Tyne.

BRIDGER, I. and LAND, N. (1992) Structured Data – An Ordnance Survey perspective, AGI Conference proceedings.

BSI (1992) British Standards Institution: BS7567 Electronic transfer of geographic information, Parts 1, 2 & 3 (NTF).

BSI (1993) British Standards Institution: BS7666, Part 1: Specification for a Street Gazetteer, Part 2: Specification for a Land and Property Gazetteer, Part 3: Specification for Addresses.

HOULT, C. and PARKER, D. (1993) The F in GIS: the use of face-based data in large scale GIS. Proceedings GISRUK conference.

PARKER, D., TAYLOR, G. E. and DAUNCEY, D. (1994) The structured concept: realising the world. *Survey Review* **32**, no. 253, 395–404.

ORDNANCE SURVEY (1993) Ordnance Survey Object Data – Specification Overview, OS Seminar, Birmingham, March 1993.

CHAPTER THREE

Balancing rights in data – elementary?

ANDREW LARNER

3.1 PREVIEW

This chapter summarises some of the research carried out for a PhD entitled, 'The legal and institutional restrictions upon the handling of digital land-related data'. The chapter starts with a review of what is meant by rights and interests in data. The chapter then examines what is meant by information and data and how well the current legal framework protects interests in data. This chapter focuses more upon the best method of protecting interests in a dataset rather than whether or not those rights are justified. It is suggested that the current legal framework is inadequate for protecting rights in data. A new method of protecting rights is proposed which is based upon the examination of the full spectrum of current intellectual property and real property law as well as the history of copyright. The chapter concludes with a brief assessment of the practicalities of implementing the proposed method of protecting rights in data.

3.2 INTRODUCTION

Land Information Systems (LISs) are used for many purposes and the data that they use are acquired from a variety of sources and processed in a number of different ways. The data used in an LIS may be referred to variously as information, data, datasets and databases.

There are a variety of *rights* in data. Hardy Ivamy (1988) describes a right as a 'lawful title or claim to anything'. Any one set of data will relate to a number of rights. Rights in a dataset might include the right to:

- use data
- stop the use of data
- create data from data
- stop the creation of data from data
- collect data

- stop the collection of data
- delete data
- stop the deletion of data
- sell data
- buy data
- look at data
- stop data being looked at
- give away data
- stop data being given away.

The rights in a dataset may belong to more than one party – for example, many parties may have the right to use a dataset. The sum total of a party's rights in a dataset may be referred to as their interest in it. The extent to which the law protects a party's interest in a dataset is dependent upon its ability to protect, or even recognise, the individual rights they have in any given set of circumstances.

While a right is a 'lawful' title or claim, some parties may try to exercise control over a dataset in the mistaken belief that it is their lawful right. This pseudo right may have the appearance of a legal claim where the legal situation is unknown to the other parties with interests in the dataset, or where the authority of the organisation is such that others are unwilling or unable to challenge it.

The mistaken belief that a right has been acquired may arise under many circumstances, including the creation and the acquisition of data. Those who have contributed to the creation of the dataset may feel that they have rights in that dataset. However, the law may consider that the contribution of a party should not give rise to a right, as is usually the case for an employee carrying out the normal duties of his job.

A party having acquired a dataset legitimately may feel that they have acquired an interest that comprises many rights. However, the licence conditions under which the dataset was acquired may not allow for all the rights that the purchaser believed that they had acquired. The purchaser of Ordnance Survey digital map data does not acquire the right to use those data without further payment.

As a dataset is created and maintained new interests in that dataset may be created – for example, a dataset may be sent to a third party for processing and the results of the processing may modify the content of the dataset. The party who carried out the modification of the dataset may now have an interest in the dataset. In some circumstances the 'modification' may be considered to be the creation of an entirely new dataset in which the supplier of the original data has no interest.

Some rights in data restrict what may be done with data and in extreme circumstances may make those data unusable. The data about fields collected for the IACS (Integrated Agricultural Control System) survey could not be re-used for the Department of the Environment land use stock survey because of personal privacy considerations.

The current legal framework has tried to re-interpret existing legal concerns in the context of computer systems. In the United Kingdom ownership of data products and personal privacy are regulated using the Copyright Designs and Patents Act 1988 and the Data Protection Act 1984 respectively. To date no consideration has been given to a unified legal framework for data. Furthermore, little consideration has been given to the nature of data and the rights in data that the law is seeking to manage. It may be that the existing Acts are not far from the required legal framework, however a judgement on the applicability of existing legislation cannot be made without an understanding of the nature of the data.

3.3   WHAT ARE DATA?

The words 'data' and 'information' are often taken to mean the same thing. However, what is actually meant by information or data is rarely specified precisely. If we cannot specify what we mean by data and information then it is unlikely that the law will ever be much help in protecting our interests in them.

We receive observations of the real world either from observing it or through communication with others. Observations may be made using the body's senses directly or using some form of instrumentation. Observations are necessarily selected, and thus the way in which they are selected will conform to the observer's expectations of the real world.

The mind of the observer records the observations of the real world or communications and quite separately builds from those a picture of reality. The separate recording and analysis by the mind of the real world was first proved by Penfield (1952) who showed that the brain was capable of playing back recordings of events and simultaneously analysing them. Thus, the meaning derived from interpreting observations and communications occurs in the mind. This process is represented in Figure 3.1 where contemplative thought is the process of analysing messages and amending existing concepts to fit newly acquired facts and concepts.

The ability to extract meaning from observations is learned. Berne (1961) argues that, in the early years of a child's life, before the development of language and social interaction, the events experienced are imposed upon the memory which form that child's basic view of the world. These events may be analysed only after the development of language and meaning in the child's mind. Once the individual has a command of language, meaning may be gathered through communication as well as through direct observations. Communication is represented by the 'message' box in Figure 3.1. Strictly speaking, the message – once in physical form – is another object; however, for emphasis, it has been drawn separately.

The study of semiotics is used to explain the communication process. Semiotics divides the analysis of a message into four levels:

- pragmatics
- semantics
- syntactics
- empirics.

The semantic level refers to the meaning of a message which Liebenau and Backhouse (1990) refer to as 'the connection that agents make between signs and their behaviour and actions'. Agents are people or things that take actions in a system. Signs are used to represent the mental model of reality that has been observed. The receiver of a message must be able to read and interpret those signs in order to build the model of reality in their own mind in order that they may react in an appropriate way. This model of reality is the 'meaning' or information that the message was intended to convey.

In order to interpret the signs in a message the connection between signs and actions must be common to both sender and receiver. Penfield discovered that

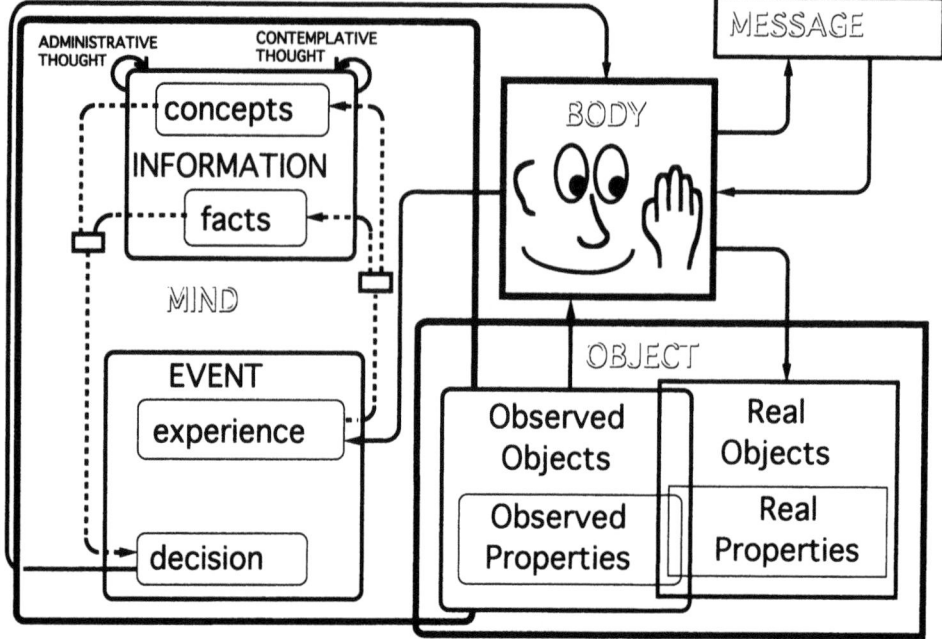

Figure 3.1 The relationship between information, observations and communications, (adapted and developed from Larner and Ralphs, 1994).

memories are locked to feelings that relate to the events observed. Searle (1986) maintained that the purpose of a message may be understood only by knowing the feelings that lie behind it. Tough (1977) demonstrated that the child learns language and meaning first through the association of emotions with sequences of events.

A child begins to understand the 'meaning' of a state and takes appropriate action. Much later the child begins to associate the state with a structured sign which is used to communicate the desired action. Thus, by sharing experiences between the adult and the child, signs are developed which are associated with facts and concepts and cognitive models of the real world are shared. These cognitive models and their underlying facts and concepts all have the use of language as signs associated with them. Thus, on future occasions, cognitive models can be communicated by the use of a shared language.

The ability to interpret the meaning in the message relies upon the ability to:

- physically receive the message
- use the same language, including grammar, vocabulary and syntax
- understand the social context of the message.

The physical transmission of a message is the empirics level of semiotics. The language used comprising the signs and their arrangement is the syntactics level. Liebenau and Backhouse categorise the signs into groups:

- an index – that relates the sign to what it signifies – for example, the greenness of a field indicating fertility
- an icon – this acts upon the senses in the same way as the thing that it refers to
- signs and symbols – have loose connections to their referent that are only established by social convention – for example, the red traffic light indicating stop.

The connection made between signs and actions at the semantic level is very dependent upon the social context. Context can include very wide social groups or a narrow 'work place' setting and can restrict the meaning of messages communicated. Pragmatics considers those aspects of thought that are common to the sender and receiver in order that they may share the meaning of a message.

In his analysis of the development of language in different cultures, Whorf (1956) identifies that the Eskimo language has three words which relate to the English word snow. A similar picture is found when looking at the Hawaiian words for lava. In each of these cultures the experience of these elements obviously plays a key role in society. In the Hawaiian case particularly so, since the names of the lava relate to the speed with which the lava flows.

The work of Penfield, Searle, Tough and Liebenau and Backhouse covers neuro-surgery, language development and communication theory. The synthesis of their research suggests that information is a cognitive model of reality. Each cognitive model is formed of concepts and records of facts. If information is to be supplied to another person who is making a decision the observer must convey this cognitive model to the mind of the decision maker. Thus, as stated by Bedard (1986): '. . . direct human communication is a cognitive-physical-cognitive modelling sequence'. Thus, a useful definition of information is: the meaning intended to be communicated to another, or the meaning created by interpreting a message or contemplating observations of the real world.

A message conveys the meaning which is built into information. The meaning is conveyed not because the message contains information but because it is inextricably linked to the concepts and records of facts that form the meaning in the mind of the receiver. From the relationship between the physical message and the cognitive models of the sender and receiver of a message we may define data as: records of fact whose interpretation results in the creation of information in the mind of the interpreter.

The description of geographical data has been a subject of much recent debate. Ralphs (1994) reviewed the description of data and proposed four 'layers' of interest, the most atomic of which is the logical geographical object. A particular logical geographical object will be described using symbols within a message and it is this object or collection of symbols which is of interest to the sender and receiver of the message. Thus, for land-related data the desired element of protection for the database creator is very small groups of symbols – for example, the corner co-ordinates of a house along with its feature code and a numerical identifier.

As in the general case described above, the context and language of the message must be apparent to sender and receiver. The sender and receiver of the data must have the same concept of reality and ways of abstracting and representing them. In the example used both sender and receiver must be interested in and have the same definition of 'house'. In addition, both sender and receiver must both abstract house as a series of lines and represent this as the co-ordinates of turning points in a database.

3.4   THE CURRENT LEGAL FRAMEWORK FOR DATA

The full balance of interests in data are not considered explicitly in any statute in the United Kingdom. In the United Kingdom the collection and use of personal data is explicitly controlled by the Data Protection Act and to a certain extent (with respect to privacy of photographs) within the Copyright, Designs and Patents Act 1988. The Copyright, Designs and Patents Act 1988 is the basis for ownership of works containing data.

The law of copyright is one of a range of legislative provisions relating to *intellectual property*. Intellectual property has no agreed definition. However, one of its distinguishing characteristics is the fact that the act of copying the property directly deprives the copyright owner of the ability to make a living out of his or her investment. The Copyright Act therefore bestows upon the copyright owner the sole right to perform certain acts, referred to as restricted acts, with respect to the work. The enforcement of restricted acts is intended to bar the purchaser from copying the work without permission.

The extent of protection afforded by copyright is restricted in two ways: infringement of copyright is not deemed to have occurred if the copying is related to a purpose which serves the interests of society or if the copying is insubstantial. These purposes are referred to as 'fair dealing' within the 1988 Act. The extent to which 'fair dealing' and 'insubstantial copying' are allowed is not specified in law.

Qualification for copyright protection in the United Kingdom is governed by the 'sweat of the brow' principle. In the Copyright Designs and Patents Act 1988 protection is extended to 'original' literary, dramatic, musical and artistic works. In UK law, originality is held to mean that the authors produced the work purely by their own effort. Some types of land-related data, such as maps, are explicitly afforded copyright protection by statute. The UK test of originality has been shown to extend to works which are collections of facts. In *Cobbet* v *Woodward*, it was held that a Post Office Directory could be subject to copyright protection.

The original copyright work is afforded protection only in so far as it is the expression of an idea. Thus, original works are only protected once they are recorded in some physical form. If the expression of the idea is dictated by the purpose of the work then the expression is said to have 'merged with the idea' and the work will not be protected. This principle is particularly strong in the United States. In the United Kingdom, while this principle is not either in statute or case law, Bainbridge (1990) argues that certain cases have been decided in a way that conforms to this principle – for example, in *Page* v *Wisden*, it was held that a cricket scoring sheet was not protected by copyright. However, from the judgement in *Cobbet* v *Woodward* it is likely that a list of the actual scores achieved in a game would be capable of copyright protection.

Certain collections of basic facts may be too commonplace for copyright protection. In *Cramp* v *Smythson*, protection was not afforded to the tables of weights and measures within a diary. Land-related data may be commonplace in that two data gatherers should come up with the same facts when collecting data for similar purposes. However, while land-related data may be commonplace within the definition used in *Cramp* v *Smythson*, the skill and care necessary to prepare for data capture and that used in deciding which facts to collect should make the land information qualify for copyright protection. In *Microsense Systems* v *Control Systems Technology*, Aldous J held that the skill and labour in devising the functions to be taken into account in the control of pelican crossings should be taken into account in determining whether the work qualified for copyright protection.

Surprisingly, in some circumstances a copyright infringement may occur even where the work does not qualify for copyright protection. In the case of *Kenrick* v *Lawrence*, copyright protection was denied to a drawing but the Judge did mention in passing that the exact duplication of that drawing would infringe copyright. In general, the strength of the protection would depend upon the value of the infringement; if there was no harm done the damages would be minimal, as was the case in *Blair* v *Osborne & Tomkins*.

In addition to directly copying collections of facts, the provider of data does not wish the user to create competing products from the supplier's data – for example, the calculation of a road centre line from the kerb lines created in a dataset. Using facts from within a work to create new facts may be considered to relate more to the idea than the expression of the idea. A similar but longer standing problem can be seen in plays where it is still not clear that the plot underlying the words is afforded protection. Bainbridge (1990) quotes the judge in the case of *Nichols* v *Universal Pictures Corporation* as saying of the boundary between idea and expression: 'Nobody has ever been able to fix that boundary, and nobody ever can'.

While works containing records of facts can be protected by copyright individual facts may not be protected by the principle of de minimis non curat lex. The de minimis principle would have it that the law should not afford copyright protection to works that are too small. By this principle words or small collections of words are not protected – for example, the quotation of a piece of a sentence was too small for protection in *Sinanide* v *La Maison Kosmeo*. If records of facts cannot be protected it is difficult to protect a work of land-related data which is a collection of facts. The crux of this problem is that in copyright, creativity should lead to easily distinguishable works allowing copying to be easily detected. In fact, works of land related data must, as seen in the previous section, have a sameness in order that they may be used.

Under intellectual property law, patent protection is used where expression of the idea leads to 'sameness'. Patent works by grant of a monopoly to the first registered inventor. This can be a greater incentive to creativity. Another advantage of patent law is that the system of registration makes pursuing infringement simpler and cheaper.

The difficulty of proving infringement and the cost of legal redress makes enforcing copyright protection difficult, with most cases settled out of court. As a result case precedent is difficult to find. The cost of taking a party infringing copyright to court is measured in tens of thousands for a simple case and hundreds of thousands for a complex one. If a point of law requires clarification then the case will almost certainly go to the House of Lords, in the case of *BLMC* v *Armstrong Patents* the total bill for costs was in excess of one million pounds.

While patents may simplify the law through registration it does have disadvantages. There can only be one winner with the award of a monopoly and the benefit of parallel exploitation of an 'information source' allowed by copyright is lost. The patent approach is to try and protect the supplier absolutely and this can stifle innovation through the fears of:

- inventing but getting beaten to registration
- the research not creating an invention.

In order to gain patent protection a work should show inventive step. A work is considered to have inventive step if a panel of experts looking at an invention do not feel that it was an obvious extension of current knowledge. The inventive step required for patent registration is not appropriate for a dataset since it cannot have too many surprises if it is to be understood.

If a legal framework to protect data is to be created then a hybrid of the patent and the copyright approach is required. This approach should allow parallel exploitation of an information source and protect unambiguously works that are the same as others. The key problem with achieving this is proving that a dataset is or is not copied.

3.5   A NEW LEGAL FRAMEWORK FOR DATA

Under the current legal framework the information sector of the economy is losing money. The requirement to collect the IACS survey data again cost millions of pounds, and losses due to illicit copying of software for the Lotus software company were estimated as $160 million per year in 1989. There are no figures for losses due to data piracy, however, its importance for land-related data is evidenced by the recent case of *Ordnance Survey* v *Mr M Younger and others*.

From the brief discussion of the nature of data and of copyright works it can be seen that currently the law does not easily afford protection to land-related data. Furthermore, the nature of data and of the failings of the law of intellectual property would suggest that any new legal framework should take account of the following:

- works may be similar to or the same as others
- data are important at the level of objects
- the interests in a dataset should be unambiguous
- the interests in a dataset should be recorded and maintained accurately
- datasets are usually subject to overlapping interests
- cost effectiveness and accessibility to the law.

The protection of works that are the same will require the creator to prove that his work has been created without 'helping himself to the product of the skill and labour of his rival' (Whitford Committee, 1977). This requires the creator to show the lineage of the dataset which could be a part of a formal system of registration of title.

It is for similar requirements for certainty in dealing with land that real property law in many countries throughout the world has moved from a system of private conveyancing to a system of registration of title. The legal community has debated whether or not copyright works are actually property (Whale and Phillips, 1983). It is clear from an examination of copyright, that in certain circumstances it resembles real property more closely than personal property. In the case of *Lauri* v *Renad*, the real property concept of 'Tenants in Common' was used to describe the combination of joint authorship interests. In the case of *Blair* v *Osborne & Tomkins*, Lord Denning MR considered that the users interests in architects plans passed with the land.

The purchaser of a real property estate held in trust must be certain that when they acquire the land that they have paid suitable consideration to the correct parties. The registration process deals with this problem. Exactly the same problem of notice occurs in copyright where beneficial ownership has been bestowed in a variety of cases. In the case of *Warner* v *Gestetner Ltd*, beneficial ownership was conferred because the formalities of assignment were not fulfilled. In addition registration would cure the title of datasets of unknown origin, as occurs with real property.

The use of registration for copyright is not new, although it has fallen out of favour in the legal community. Indeed, its use is overlooked in many legal texts making a thorough investigation of the reasons for its use and its demise difficult.

The printing press was introduced into England in 1476. The first formal use of copyright registers was in 1556 as a result of the Star Chamber Decree. The first copyright statute was not enacted until 1709 (Statute of Anne). Copyright registers finally disappeared when the United Kingdom became a member of the Berne Copyright Union in 1887. Thus the use of copyright registers runs for some 300 years, or longer if the United States experience is taken into account.

While the use of registers has a long history the current approach is much more recent. The principle of copyright arising at creation only became formally accepted in the 1709 Statute of Anne. However, between 1709 and 1887 it was still necessary to register in order to claim infringement in Court (*Beckford* v *Hood*). It is only since 1887 that the current copyright regime started to develop.

Whale and Phillips (1983) argue that the register approach was abandoned because its underlying use was as an instrument of censorship. Once the principle of free flow of ideas took root, the registers became unimportant to government. Another adverse factor of the register was that it registered only certain rights – for example, author's rights were not registered. These factors do not militate against the re-introduction of registration since the basis for the new registration system would be the creator's desire for certainty of title and the user's desire for certainty of dealing with the dataset.

The willingness of those involved with data to create very detailed registers and the ability to manage these dynamically has become very apparent in recent years. The need for metadata, data describing a dataset, has led to a number of data documentation standards. Furthermore, there are many attempts world-wide to register government datasets, including the register of the United Kingdom government's Tradable Information Initiative (Larner, 1992). In 1994 a database describing datasets was set up by European Mapping agencies. This database (the GDD) is physically located in Frankfurt but is accessible anywhere in the world with a computer and telematic link. The development of the Internet as a communication and data delivery link makes a dynamic register or network of registers a practical possibility.

3.6 A MIRROR OF ALL RIGHTS IN A DATASET

In real property law the mirror principle maintains that the register should accurately reflect all the interests in the land. If a data register is to reflect accurately all interests in a dataset then the interests described will go outside of the traditional copyright arena. Larner (1994) shows that land-related data may be personal in many circumstances. Registering land-related datasets will require the registration of the data subject's interests to conform to the mirror principle.

While the connection between personal privacy and ownership rights in data has not been made explicitly by the legal or GIS communities there are some clear connections. In the 1988 Copyright Act, by section 85 there is a right to privacy of photographs and films commissioned for private and domestic purposes. While new to statute law this principle is in sympathy with the approach taken in using economic rights to ensure privacy as with *Mail Newspapers plc* v *Express Newspapers plc*.

In other countries there has been debate about the data subject's ownership of personal data. In France, a couple who were 'caught' in a photograph which subsequently became famous and earned substantial money tried to claim ownership of the photograph. In the United States ownership of unique gene codes was given to the person whose body they formed a part of. Even if the rights in data that arise from personal information are not considered ownership rights they do operate in the same way as burdens in real property law. These burdens are in effect a part of the existing register run by the Office of the Data Protection Registrar.

As with the data rights that arise from creating data, personal data rights operate at the level of objects in the database. The Data Protection Act 1984 gives the data subject rights over data including rights of:

- access
- correction
- use
- deletion.

3.7 SUMMARY

While the law in the United Kingdom does not currently go as far as protecting individual records of facts, its approach to protecting the copyright may still be considered to be a property right.

Rights in data bear some similarities to rights in land. An effective system of registration of title to data would have the same benefits in freeing up the commercial data market as the registration of title to land has had to the real property market.

Registration of title to data would form the basis for the creation of statutory licences to produce value added data products. In order to achieve this the registry would have to operate in the same efficient way in which the land registry takes on titles, including the ability to cure defects. The basis for this activity already exists in the Copyright Tribunal and in the Data Protection Tribunal. In addition there would have to be a level of insurance operating to cover mistakes by the registry.

There are some indications that the land-related data community favour registration of datasets for a variety of reasons. Rights arising from personal data are already part of a system of registration. However, if a complete system of registration of title to data can be operated it may be necessary to change the Berne Copyright Union's opposition to the necessity of formalities for the qualification for copyright protection. Alternatively, since the bona fide purchaser of copyright works takes free of any interests in a work, registration may have a formal role as a source of actual, constructive or implied notice.

REFERENCES

BAINBRIDGE, D. I. (1990) *Intellectual Property*. London: Pitman Publishing.

BEDARD, Y. (1986) A study of the nature of data using a communication-based conceptual framework of land information systems. *Canadian Surveyor*, **40**, 449–460.

BERNE, E. (1961) *Transactional Analysis in Psychotherapy*. New York: Grove Press.

BLUMENTHAL, S. C. (1969) *Management Information Systems*. Englewood Cliffs, NJ: Prentice Hall.

HARDY IVAMY, E. R. (1988) *Mozley and Whiteley's Law Dictionary*, 10th edn, London: Butterworths.

LARNER, A. G. (1992) Access to Information in the United Kingdom and the role of the AGI in the Tradable Information Initiative. Proceedings of the 2e Forum International De L'Instrumentation & De L'Information Geographique.

LARNER, A. G. (1994) Personal data protection and the Chartered Surveyor. *Chartered Surveyor Monthly*, February 1994, Builder Group, London.

LARNER, A. G. and RALPHS, M. P. (1994) Walking through the mindfield. Domesday 2000 working paper No. 2, University of East London.

LIEBENAU, J. and BACKHOUSE, J. (1990) *Understanding information*. London: Macmillan.

RALPHS, M. P. (1994) Towards the evaluation and management of spatial data quality. Unpublished PhD thesis, University of East London, London.

PENFIELD, W. (1952) Memory mechanisms. *A.M.A. Arch. Neurol. Psych.* **67**, 178–198.

SEARLE, J. R. (1986) *Minds, Brains and Science*. Cambridge, MA: Harvard University Press.

TOUGH, J. (1977) *The Development of Meaning*. London: Unwin Education Books.

UNITED KINGDOM, LAWS, STATUTES etc., 'THE COPYRIGHT, DESIGNS and PATENTS ACT' (1988) London: Her Majesty's Stationary Office.

UNITED KINGDOM, LAWS, STATUTES etc., 'THE DATA PROTECTION ACT' (1984) London: Her Majesty's Stationary Office.

WHALE, R. F. and PHILLIPS, J. J. (1983) *Whale on Copyright*. Oxford: ESC Publishing Ltd.

WHITFORD COMMITTEE (1977) Copyright and Designs Law – Report of the Committee to consider the Law on Designs and Copyright, March 1977 Cmnd 6732.

WHORF, L. (1956) *Language, Thought and Reality*, CARROL, J. B. (ed.), Massachusetts Institute of Technology, Cambridge, MA.

LIST OF CASES

	Section (S) and paragraph (P) reference
Beckford v *Hood* (1798)	S5 P7
Blair v *Osborne & Tomkins* (1971) 2WLR 503	S4 P7
Cobbet v *Woodward* (1872) 14 Eq 407 LR	S4 P4
Cramp v *Smythson* (1944) AC 329	S4 P6
Kenrick v *Lawrence* (1890) 25 QBD 99	S4 P7
Lauri v *Renad* (1892) 3 Ch 402	S5 P3
Mail Newspaper plc v *Express Newspapers plc* (1987) FSR 90	S6 P2
Microsense Systems v *Control Systems Technology* (unreported) 17 July 1991 (Chancery Division)	S4 P6
Nichols v *Universal Pictures Corporation* (1930) 1945 F.2d 119	S4 P8
Ordnance Survey v *Mr M Younger and others* (unreported) 10 April 1995 (Chancery Division)	S5 P1
Sinanide v *La Maison Kosmeo* (1928) 139 LT 365	S4 P9
Warner v *Gestetner Ltd* (1988) EIPR D-89	S5 P4

CHAPTER FOUR

Georeferencing people and places: a comparison of detailed datasets

DAVID MARTIN and GARY HIGGS

4.1 INTRODUCTION

The indirect use of postcodes or census geography has for a long time provided the only widely used means of georeferencing population and socio-economic data within Britain. However, many organizations, including local government, have developed extensive databases which contain detailed address-referenced information, and a number of unique property reference numbering (UPRN) schemes are in existence. In addition, many commercial organizations have constructed personal databases which include full address information, although the original applications for which they were created were non-geographical (customer databases, patient lists etc.). The last decade has seen a massive increase in the development of GIS applications, and a major factor in this growth has been applications which are in some way concerned with socio-economic information (Martin, 1991; Rhind, 1991). Various commentators have suggested that the availability of accurate data is both an important driving force in the implementation of new GIS, and also likely to be one of the long-term limiting factors for effective GIS use (Fisher, 1991). The 1991 Census of Population has had a significant impact on the growth of such applications, being the first census to take place within an environment of widespread GIS use and awareness. This development has been accompanied by a corresponding increase in the market for georeferenced data and value-added products. The objects of ultimate interest in socio-economic applications are usually individual persons or households, but it remains very difficult to reach clear and workable definitions of an individual's location and, if obtainable, such information would present a number of problems associated with the protection of privacy.

In the following section, the development of increasingly detailed georeferencing systems in Britain is reviewed, with particular reference to the importance of property addresses, culminating in the adoption of a new British Standard for address referencing and the release of ADDRESS-POINT™. The existence of these many different scales of georeferencing and data publication pose important problems for the integration of existing datasets which may be required both in commercial and research applications. We illustrate the many different ways in which the existing datasets may be manipulated within standard GIS software in order to provide 'hybrid' geographical references, and move on to address the apparently simple task of defining a small area and giving its population. There is a need to understand the implications of these different systems for GIS use, and we draw on an empirical comparison of some of the current georeferencing options for the City of Cardiff in South Wales. This work has arisen out of substantive research which was concerned with the examination of changing local taxation regimes, and which utilized a variety of existing urban databases (Longley *et al.*, 1993; Martin *et al.*, 1994).

4.2 INCREASINGLY DETAILED DATASETS

Digital data products for georeferencing fall into three main types: these are data-sets relating to census geographies, datasets relating to postal geographies, and datasets structured around address lists. In addition, they comprise a variety of boundary data, centroid data and point data. The Chorley Report (Department of the Environment, 1987) recognized the importance of property addresses as a basic building block in the creation of geographies for data relating to population, and suggested the use of postcode geographies as a suitable georeferencing system. Recent development of socio-economic GIS has made heavy use of postcode systems, but even here there are problems associated with the precise definition of postcode locations. Also available since the early 1980s have been the central postcode direc-tory (CPD) and postcode address file (PAF), created by the Post Office. The main method for locating addresses was then (and for many applications has remained) via the CPD, which provides a single national grid coordinate for each unit post-code (the smallest division of the postcode system, typically containing around 15 addresses). The file was not originally created with detailed address georeferencing in mind, and fears about the accuracy of its 100 m grid references have persisted. The PAF is the Post Office's definitive listing of all postal delivery addresses, of which there are approximately 24 million, and contains the full text of each address, together with the correct postcode. This list is constantly updated as properties are constructed and demolished, and provides a route through which any postal address may be allocated to its correct postcode and thus to a 100 m grid reference from the CPD. These datasets have grown in importance with the continued increase in the use of postcodes as general purpose georeferencing codes (Raper *et al.*, 1992).

A number of organizations hold property-level databases which are not explicitly georeferenced, but which contain addresses in some form. Examples would include the rates registers maintained by local government and water companies, the regis-ter of electors and the council tax register (valuation list). Such registers frequently contain some form of unique property reference number (UPRN), which identifies each address or subdivision of an address, such as the flats or bedsits in a large subdivided house. UPRNs may allow the organization to manage billing or mailing operations, but do not usually contain any explicitly spatial component such as a grid reference. So-called 'intelligent' UPRNs may be constructed which incorporate components describing (for example) approximate location and property type. Due to the specific purpose for which they were created, each of the different registers use different definitions of properties, and suffer different problems of omission and sys-tematic error.

An important development in the late 1980s was the development of a new product by Pinpoint Analysis Ltd, known as the Pinpoint Address Code (PAC). This product contained a single grid reference for each property, linked to the Postcode Address File. The advantages of such a high-resolution dataset were clearly demonstrated by Gatrell (1989) and Gatrell *et al.* (1991), who compared the performance of the CPD and PAC data in distance-based and point-in-polygon assignments of properties to census enumeration districts (EDs) and wards. Many new tasks were possible using the address-level data. For various reasons PAC failed to become the national property-level referencing system which was so clearly needed. The original PAC data (partial national coverage, concentrated in major urban areas) are still commercially available as Grid Point.

For many users, linkage of postcoded data with the Census of Population is an important application, which is made additionally complex by the fact that there is no correspondence between census and postcode boundaries in England and Wales. (In Scotland, by contrast, the postcode system was used as the basis for the design of census geography in 1981 and 1991.) This situation was greatly aided by the production of a new ED to postcode directory following the 1991 Census (Martin, 1992). The directory provides the CPD postcode grid reference together with a count of the number of households falling within each unique postcode/ED combination. These new units defined by the intersection of postcodes and EDs are known as part postcode units (PPUs), and for each postcode the ED in which the majority of its population falls is also given, being referred to as pseudo-EDs (PEDs). A wide range of statistical information, together with digital boundaries and a single centroid location are available for each ED. Fuller details of the various 1991 Census data products will be found in Dale and Marsh (1993) or Martin (1993).

It is against this background that ADDRESS-POINT was launched in April 1993 (Ordnance Survey, 1993). ADDRESS-POINT has been initially created by matching the PAF against Ordnance Survey's Land-Line digital database, using OSCAR (road centreline) thoroughfare information as a guide. In most cases the property seed point from Land-Line is then assigned as the address location, with a resolution of 0.1 m. In addition, a unique Ordnance Survey ADDRESS-POINT Reference (OSAPR) is given to each address. The ADDRESS-POINT data contain status flags which indicate the seed point type, positional quality, physical status and match status of the address reference. ADDRESS-POINT thus provides a direct solution to a number of the difficulties with earlier, and lower resolution, property georeferencing methods.

Since the early 1980s we have seen the most detailed of the available basic spatial units shift in scale from the small census zone, down through the postcode, and finally resting with the property address. The emergence of similar situations may be seen in many countries, with widespread use of postal delivery code systems, census geographies and property databases as the basis for GIS, and the growth of associated value-added data products. Elsewhere, as in Britain, digital data associated with the censuses of the early 1990s have provided an important spur to commercial exploitation and academic research in this field.

In order to achieve linkage between any of the datasets discussed here, it is necessary to identify common keys. For many applications, the postcode has been the common key which is present in each data series and which has allowed georeferencing. However, as we move increasingly towards the use of individual

Table 4.1 Georeferencing options at the sub-ED level

Geographical unit	Point reference	Boundary data
Property	ADDRESS-POINT (A-P)	Land-Line *Thiessen around A-P*
Part Postcode Unit (PPU)	*Weighted mean from A-P*	*Thiessen around PPU point references* *Merged Thiessens around property points*
Unit postcode	*Weighted mean from A-P* *Weighted mean from postcode/ED directory (PCED)*	*Thiessen around postcode point references* *Merged Thiessen around property points*
Enumeration district (ED)	Given in Small Area Statistics (SAS) *Weighted mean from PCED* *Weighted mean from A-P*	ED-line ED91

property-level records, the address itself must become the common key. Our work on local government registers in Cardiff, described elsewhere (Martin *et al.*, 1994) has involved the linkage of individual property records across a number of different property-level databases, and this has clearly demonstrated the need for consistent address referencing, if existing databases are to be fully integrated. For these reasons, there has been increasing pressure for the adoption of a national standard address format, and the assignment of a single unique UPRN to each identifiable property in Britain. This has culminated in the establishment of British Standard 7666 parts 2 and 3 in June 1994. These specify a standard format for address and property referencing (Cushnie, 1994). BS7666 part 2 provides the basis for a national land and property gazetteer. Such a gazetteer would be an index for all Basic Land and Property Units (BLPUs), where these are contiguous areas of land under uniform ownership rights. The use of a definition based on ownership was eventually adopted as this is the only definition of a land parcel which is documented and therefore consistently reproducible (Green, 1993). The third part of BS7666 specifies a standard address structure. This is intended to provide a consistent format for all addressable objects.

4.3 HYBRID GEOREFERENCING

It is apparent that many new data products may be created by the combination of two or more of the datasets described above. For some applications, the 'hybrid' products may offer the most cost-effective mechanisms for the representation of socioeconomic information, or for the integration of two datasets originally published for incompatible areal units, and some of these possibilities are evaluated in our empirical study below. In this paper, we are concerned solely with the use of existing databases, and have explicitly chosen not to focus on the potential alternatives such as the population surface models described by Bracken and Martin (1995).

Table 4.1 illustrates the different datasets which may be used to represent the four different levels of sub-ED geographical referencing (i.e. EDs, postcodes, PPUs and properties). Considerable variety exists in the methods which may be used for the creation of new point and boundary references, but new point layers may in general be derived either by weighted spatial means of references for lower level units, or by calculating geometric centroids from boundary data. New boundary data may be created by generating Thiessen polygons around appropriate point references, by merging lower level polygons, and where necessary clipping to higher level boundaries. Each of the italicized datasets in the table represents a hybrid georeference, created from one or more given datasets, and it is suggested that those derived from the more detailed products such as ADDRESS-POINT will in fact offer greater locational accuracy than those which are currently available. For example, the spatial mean of the ADDRESS-POINT references falling within an ED will be a more useful centroid than that given in the census SAS. In other cases, such as the boundaries of PPUs, the hybrid data are in fact entirely new, as there are currently no boundary data available (or defined) for these polygons.

4.4 THE CARDIFF STUDY

The study area used for this paper is that which formed the basis for an ESRC-funded study on evaluation of changing local taxation regimes (Longley *et al.*, 1993), and comprises around 47 000 properties in the Cardiff 'Inner Area'. The study area is entirely urban, comprising a large proportion of terraced housing dating from around 1900, some local authority-built housing estates, and a smaller amount of prestigious new housing associated with the Cardiff Bay redevelopment scheme. The area includes the commercial centre of the city, with many older multi-occupied properties subdivided into flats and bedsits. This study is based around an exploration of the various products identified in Table 4.1. We have sought (i) to match ADDRESS-POINT records with other existing property registers, (ii) to examine the locational accuracy both of the existing and the derived datasets as compared to ADDRESS-POINT, and (iii) to compare the estimates of small area population/household counts which are available via the different methods.

4.4.1 Matching with existing property registers

The first task which we undertook was to attempt to match the ADDRESS-POINT data with the pre-existing property registers for the same area. If existing local government databases and commercial customer lists are to be correctly georeferenced, successful list matching of this type is essential. Here, we have sought to match the postcoded council tax register with the ADDRESS-POINT data. It should be stressed that no information on council tax bandings was involved in this process: we are simply seeking to determine the degree of success with which the address lists may be integrated. The register is here used as an example of a current address list with UPRNs, in order to ascertain the extent to which georeferencing may be

achieved. Matching of registers on the basis of address text strings is highly problematic due to the wide variety of formats and address conventions which are used
as administrative datasets are created and updated by clerical staff. Enormous
variety of address formats and street and property naming exists within the rates,
council tax and ADDRESS-POINT data, making direct text matching prohibitively
complex. We therefore sought to match the registers using the postcode as a first
level filter. For each unit postcode, addresses on each register were compared in
order to seek matching properties.

A variety of incompatibilities exist between the two registers, which are typical of
those encountered when matching any large address databases: 1.3 per cent of the
addresses in the council tax register were not postcoded, and postcodes were added
to these records manually wherever possible. After this stage, records failed to match
for one of the following reasons: addresses appearing in only one of the databases;
addresses with wrong postcodes; addresses inconsistently referenced by property
name and street number; properties with composite numbers (e.g. '213–215' or '213/
215'), and incompatible address formats. In all, we were able to resolve matches for
92 per cent of the council tax addresses. The detailed process is described in Higgs
and Martin (1995), and the 'solution' still incorporates a number of one-to-many,
many-to-one and many-to-many relationships, with only 78 per cent as 'perfect'
one-to-one matches. The degree of success achieved here is comparable with that
noted in the Bristol trials of the National Land Information System (NLIS)
(Hemmings, 1995).

This study serves to reinforce our endorsement of the need for a widely used
standard address format, as defined by BS7666. The adoption of such a standard is
an essential prerequisite for the effective matching of existing registers with
ADDRESS-POINT in order to take advantage of the highest resolution grid referencing. The effectiveness of a national system of high resolution property geor-
eferences is significantly restricted when contemporary administrative databases
display significant mismatches.

4.4.2 Locational accuracy

Each ADDRESS-POINT record contains a four-digit seed point status flag which
indicates the positional quality of the geographical reference. For the purposes of
evaluating various other data products and 'hybrid' solutions, we have removed
those property records which are not coded '0354' (mostly non-permanent
addresses, or having only a 100 m grid reference derived from the CPD). The
remaining ADDRESS-POINT records (92.2 per cent) have been taken to represent
the 'definitive' locations of inner Cardiff properties.

It is apparent from Table 4.1 that there are a number of different options for the
definition of each of the spatial references commonly encountered between the ED
and property levels. In some cases, such as ADDRESS-POINT or digital ED
boundaries, we would expect that the commercially available product is of higher
quality than anything which might be created from the intersection of other layers,
but this is not the case for ED and postcode centroids, and for some of the cells in
the table there is no commercially available dataset. The high precision of
ADDRESS-POINT georeferencing allows us to use a point-in-polygon approach to

the definitive allocation of each property into its correct ED, and by checking against the postcode in the ADDRESS-POINT record, it may further be allocated into the correct postcode and PPU. This knowledge has permitted us to create a number of 'definitive' locations which may be compared with the existing products. It is also possible to derive 'improved' ED centroids by combining the postcode/ED directory and SAS information, in order to give a weighted spatial mean for each ED, based on the postcodes it contains, and the proportion of its properties that fall within it. If such a location is indeed an improvement over the SAS location, this would be an attractive low-cost route to the enhancement of the SAS centroids. With respect to boundary generation, any of the centroid datasets provides a starting point for Thiessen boundary generation where no boundaries are available, and these may be significantly improved by 'clipping' to an existing higher level boundary.

Figure 4.1 compares the locational accuracy of ED centroids derived from the SAS with those computed as a spatial mean of the ADDRESS-POINT locations in each ED. The mean error of the SAS centroids is 66.3 m with a standard deviation of 61.8 m. New ED centroids may also be derived by a household-weighted mean of the postcode locations falling in each ED, and when compared to the ADDRESS-POINT centroids, these display a mean error of 56.9 m with a standard deviation of 61.3 m. This suggests that the weighted postcode mean is a slight improvement over the SAS locations, and we anticipate that this would be more marked in suburban and rural locations.

Figure 4.2 illustrates the geographical distribution of distances between matching centroids with identical ED codes derived from ADDRESS-POINT and SAS. The largest distances are associated with larger EDs and in particular areas of the city with large numbers of non-domestic properties. In such instances it may be that the SAS centroids more closely represent the distribution of population associated with residential properties. This is also reflected in the fact that the SAS population-weighted centroids for the civic and shopping centre of Cardiff depart significantly

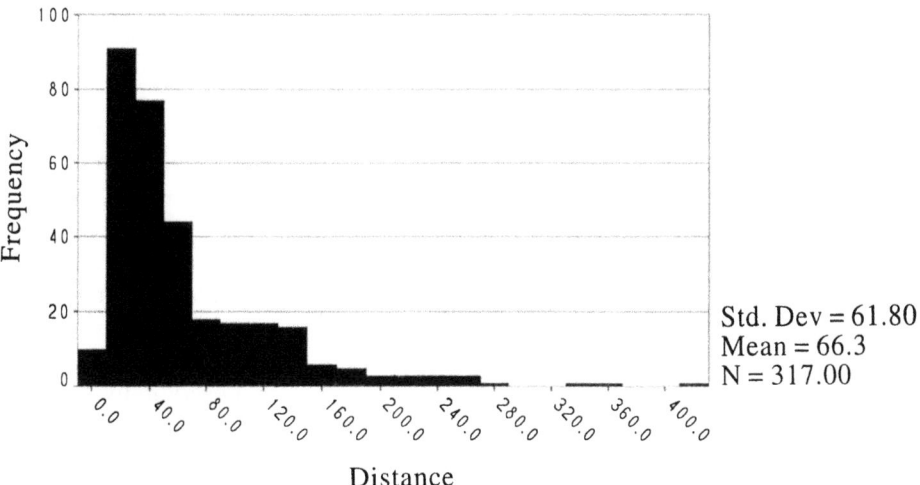

Figure 4.1   SAS and ADDRESS-POINT ED centroids compared.

Figure 4.2   SAS and ADDRESS-POINT ED centroid displacement.

from those calculated from **ADDRESS-POINT** since we have no way of distinguishing between residential and non-residential addresses. Conversely, in areas
dominated by residential properties local knowledge suggests that a weighted centroid based on **ADDRESS-POINT** is likely to be a more accurate representation of
the population distribution. It may be possible to compensate for some of these
difficulties by using additional information from the PAF about property type
(commercial, residential, subdivided properties) in order to filter the **ADDRESS-
POINT** data before computing centroids, but this has not been attempted here.

Boundaries may in theory be created for PPUs and unit postcodes by creating
Thiessen polygons around point references for these polygons, and clipping to ED
boundaries where appropriate. Unfortunately, the 100 m referencing of the CPD
and postcode/ED directory does not allow such a technique to distinguish between
the locations of all individual postcodes, and the resulting polygons may therefore
apply to one or more postcodes of PPUs. More useful polygons may be created
around the **ADDRESS-POINT** derived centroids for these zones, as these do indeed
provide definitive unique polygons which are free from the grid structure caused by
100 m referencing, while remaining clipped to the appropriate ED boundaries.
Problems arise with all Thiessen-based approaches if the underlying units are non-
contiguous, as these structures cannot be represented in the resulting polygons. A
final approach is to merge Thiessen polygons created around every individual property by dissolving the boundaries between polygons sharing the same postcode or
PPU identifier. This imposes a heavy computational burden, produces irregular
boundaries, and is again subject to problems where the underlying data are non-
contiguous. Ideally, these most detailed polygons should be integrated with physical
features such as road centrelines in order to achieve definitive boundaries.

4.4.3   Comparison of household counts

For many purposes, it is necessary to define base populations for small areas, most obviously for the calculation of incidence rates and percentages. In survey design, it may also be required to know the number of households falling in a known area, and household or property numbers are sometimes used as proxy population measures where reliable person counts cannot be obtained. The datasets used here offer a number of different estimates of 'population' in these terms. Only the census SAS give numbers of persons, but the SAS and postcode/ED directory give numbers of households (census definition), and ADDRESS-POINT gives numbers of properties (postal address definition).

Figure 4.3 illustrates the scatter of household counts at the ED level between those calculated using ADDRESS-POINT and the 1991 Census of Population. The correlation between the two counts is $r^2 = 0.545$ (and 0.49 between ADDRESS-POINT and the household counts from the postcode/ED directory). ADDRESS-POINT is likely to predict higher household counts in areas which have been recently developed (i.e. post-1991 Census). Household counts from the census are also underestimated for areas of the city where there are large tracts of non-domestic properties, including the city centre itself. Indeed, the largest discrepancy between household counts from the census and ADDRESS-POINT is in the Castle ward in the civic centre of Cardiff which is predominantly occupied by government and university buildings. Another contributory factor may be that the census has under predicted counts as a result of under-enumeration.

The census tends to over-predict the number of households relative to ADDRESS-POINT in areas of the city dominated by HMOs where the number of households recorded by the census exceeds the level of property subdivision captured by ADDRESS-POINT. Previous research suggests that this phenomenon is likely to be significant in areas with higher than average proportions of ethnic minorities and students. This explanation would apply particularly in inner neighbourhoods such as Riverside and Cathays. Further analysis (not presented here) has

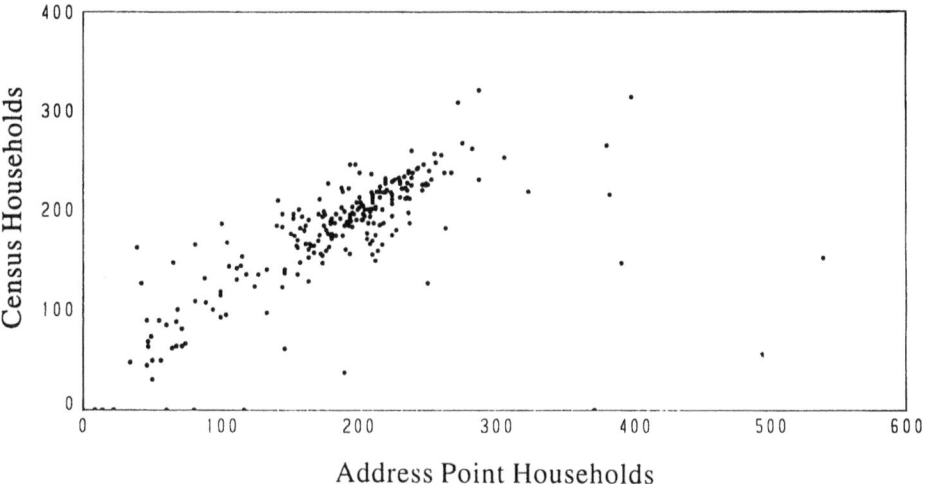

Figure 4.3   SAS and ADDRESS-POINT household counts by ED.

revealed similar variations in household counts when comparing the postcode/ED directory with ADDRESS-POINT.

4.5 CONCLUSIONS

We have reviewed the increasing geographical resolution with which it has become possible to georeference population. While increased resolution in digital geographical data is to be welcomed, it is not without its problems. One of the main difficulties concerns the adoption of appropriate standards. The need for standards in address referencing is amply illustrated by the difficulties of matching existing address registers, and in this context, the implementation of BS7666 is to be welcomed. Increased availability of high-resolution grid referencing also makes possible the creation of hybrid datasets by organizations where a definitive dataset is not held or does not exist. For a specific application, this approach may be attractive in terms of data cost and organization. However, there is a danger that the variety of alternative strategies will lead to further problems of data incompatibility in the future. As organizations are faced with the tasks of estimating socio-economic characteristics for small areas or attempting to update census data in line with residential development, there will be a temptation to take the least-cost route, even where this may result in lower quality locational data. The work presented here has illustrated some of these potential difficulties, including the nature of incompatibilities between census and address-based proxies for household counts, and the distinct geographical patterning which may be present. Awareness of these issues will be particularly important when, for example, these new datasets are used as the basis for the design of a census geography for 2001.

ACKNOWLEDGEMENTS

The authors acknowledge the support of ESRC award R000234707, and the assistance of Paul Longley, Scott Orford and Ed Thomas in the initial stages of this work, and are extremely grateful to Ordnance Survey for use of version 1.0 of the ADDRESS-POINT data for Cardiff, and to Cardiff City Council for the use of address registers for the city. All views and opinions expressed in this paper remain the sole responsibility of the authors.

REFERENCES

BRACKEN, I. and MARTIN, D. (1995) Linkage of the 1981 and 1991 Censuses using surface modelling concepts. *Environ. Plan. A* **27**, 379–390.

CUSHNIE, J. (1994) A British standard is published. *Mapping Awareness* **8**, 40–43.

DALE, A. and MARSH, C. (1993) *The 1991 Census user's guide.* London: HMSO.

DEPARTMENT OF THE ENVIRONMENT (1987) *Handling Geographic Information: the report of the Committee of Enquiry chaired by Lord Chorley.* London: HMSO.

FISHER, P. F. (1991) Spatial data sources and data problems. In MAGUIRE, D. J., GOODCHILD, M. F. and RHIND, D. W. (eds) *Geographical Information Systems: Principles and Applications,* vol 1, pp. 175–189. Harlow: Longman.

GATRELL, A. C. (1989) On the spatial representation and accuracy of address-based data in the United Kingdom. *Intern. J. Geogr. Inform. Syst.* **3**, 335–348.

GATRELL, A. C., DUNN, C. E. and BOYLE, P. J. (1991) The relative utility of the Central Postcode Directory and Pinpoint Address Code in applications of geographical information systems. *Environ. Plan. A* **23**, 1447–1458.

GREEN, R. (1993) The National Land and Property Gazetteer – the solution? *Map. Aware.* **7**, 46–48.

HEMMINGS, D. (1995) The NLIS prototype including the Bristol City Council trial of BS7666. In *Addressing Your Information*. London: AGI.

HIGGS, G. and MARTIN, D. (1995) *A Comparison of Recent Spatial Referencing Approaches in Planning*. Papers in Planning Research 155, Cardiff: Department of City and Regional Planning, UWCC.

LONGLEY, P., MARTIN, D. and HIGGS, G. (1993) The geographical implications of changing local taxation regimes. *Trans. Inst. Br. Geogr.* **18**, 86–101.

MARTIN, D. (1991) *Geographic Information Systems and their Socioeconomic Applications.* London: Routledge.

MARTIN, D. (1992) Postcodes and the 1991 Census of Population: issues, problems and prospects. *Trans. Inst. Br. Geogr.* **17**, 350–357.

MARTIN, D. (1993) *The 1991 UK Census of Population.* Concepts and Techniques in Modern Geography 56. Norwich: Environmental Publications.

MARTIN, D., LONGLEY, P. and HIGGS, G. (1994) The use of GIS in the analysis of diverse urban databases. *Comput. Environ. Urban Syst.* **18**, 55–66.

ORDNANCE SURVEY (1993) *ADDRESS-POINT User Guide.* Southampton: Ordnance Survey.

RAPER, J., RHIND, D. and SHEPHERD, J. (1992) *Postcodes: the New Geography.* Harlow: Longman.

RHIND, D. (1991) Counting the people. In MAGUIRE, D. J., GOODCHILD, M. F. and RHIND, D. W, eds, *Geographical Information Systems: Principles and Applications*, vol 1, pp. 127–37. Harlow: Longman.

CHAPTER FIVE

The assessment of catchment environmental characteristics and their uncertainty

DAVID MILLER, RICHARD ASPINALL, JANE MORRICE, GARY WRIGHT and ALLAN LILLY

5.1 INTRODUCTION

The use of hydrological catchments as a basis for environmental research is related to their functional role in integrating chemical, biological and physical processes (Band, 1989; Moore et al., 1991). The extensive modelling of hydrological processes now being undertaken using GIS is increasingly taking advantage of spatial analysis, data management and display and visualization functionality (Maidment, 1993; Ferrier et al., 1990). Many hydrological models operate on catchments as a lumped entity. Catchment characteristics are simplified to single values and ignore within catchment variability. These catchment attributes are dependent on the reliability of delimiting the catchment and on the accuracy of summarizing attributes within the catchment. This paper reports on the issues of using such lumped data in the modelling of catchments. An analysis of the consistency and reliability of catchment delimitation based upon topographic maps of different scale is presented for selected catchments.

Lessons from this study were used to structure a database of standing water catchments in Scotland, described in terms of their land cover and soils. The use of catchments as a basis for describing and discussing the implications of land use on the environment depends, in part, however, not only on catchment boundaries but also on the reliability of land cover descriptions for each catchment. Validation data for land cover (Aspinall and Pearson, 1993) is used to rank the catchments according to the reliability of land cover interpretation for each catchment. This adds to the information that can be used to produce hydrological model output.

5.2 CATCHMENT DELIMITATION

The uncertainty associated with manual interpretation of catchment boundaries from topographic maps has been assessed for selected catchments in the Trossachs area of Central Scotland. The terrain of this area is very varied and graphical delimitation of catchment boundaries and the comparison of these interpretations, has been undertaken from a range of map scales available for Great Britain. The map scales have been chosen because of their use in environmental applications such as runoff modelling, waste disposal planning, habitat mapping (Towers, 1994; Aspinall, 1994) and wide availability to the modelling and GIS communities.

Three scales of topographic maps were used in this study: 1 : 10 000, 1 : 25 000 and 1 : 50 000. Each of the catchments for the eleven lochs were delimited from these topographic maps. The lochs, range in size from Lochan Reoidhte (0.02 km^2) to Loch Katrine (13.27 km^2). Table 5.1 is a summary of the lochs, and Figure 5.1 shows the catchment boundaries as interpreted. Of these lochs, seven were surveyed in the 1895 bathymetric survey of Scottish Freshwater lochs by Murray and Puller (1900) who provide surface area and depth information plus an estimate of catchment area. The catchment areas given were based upon the topographic maps available at the end of the last century.

Comparisons of the boundaries from 1 : 50 000 maps show that the maximum perpendicular distance between interpreted boundaries was up to 780 m (at ground scale). The total perimeter of catchment boundaries in this area is 168 km. For less than 5 per cent of the catchment boundaries the differences between each boundary is greater than 200 m apart, whereas for over 70 per cent of the perimeter the difference is less than 25 m. Thus for many parts of the catchment perimeter, the accuracy of positioning approaches the accuracy quoted for the map (that is, 90 per cent of all contour lines are within = ± 0.5 m of the contour interval of 10 m (Ordnance Survey, 1992)). This varies however, by catchment size and for the four smallest catchments the percentage of the perimeter that varies by over 200 m is up to 35 per cent (e.g. Loch Drunkie). Overall, the boundaries derived from the 1 : 10 000 and 1 : 25 000 scale map data follow each other more closely compared to the boundary from the 1 : 50 000 scale data. The difference between the boundaries is less than 25 m for over 85 per cent of the length.

Table 5.1 Lochs and catchment areas derived from different scales of data (na = not available)

Name	Area (km^2)	Map source scale			Murray and Pullar
		1 : 10 000	1 : 25 000	1 : 50 000	
Loch Katrine	13.27	94.11	94.11	94.43	96.92
Loch Vennachar	3.88	33.80	34.93	34.75	74.20*
Loch Arklet	2.24	19.03	18.07	18.15	13.80
Loch Ard	2.22	21.51	21.75	21.91	na
Glen Finglas Res.	1.41	39.89	38.83	39.17	74.20*
Loch Chon	1.09	15.59	15.35	19.16	na
Loch Achray	0.74	19.49	19.38	19.16	18.13
Loch Drunkie	0.59	6.01	4.92	4.95	5.7
Loch Tinker	0.12	1.12	1.30	1.19	na
Lochan Mhaim an Carn	0.03	0.44	0.42	0.42	na
Lochan Reoidhte	0.02	0.31	0.17	0.16	na

* In 1900, Loch Vennachar comprised the sum of both the catchment delimited from maps dated 1985 plus that of Glen Finglas.

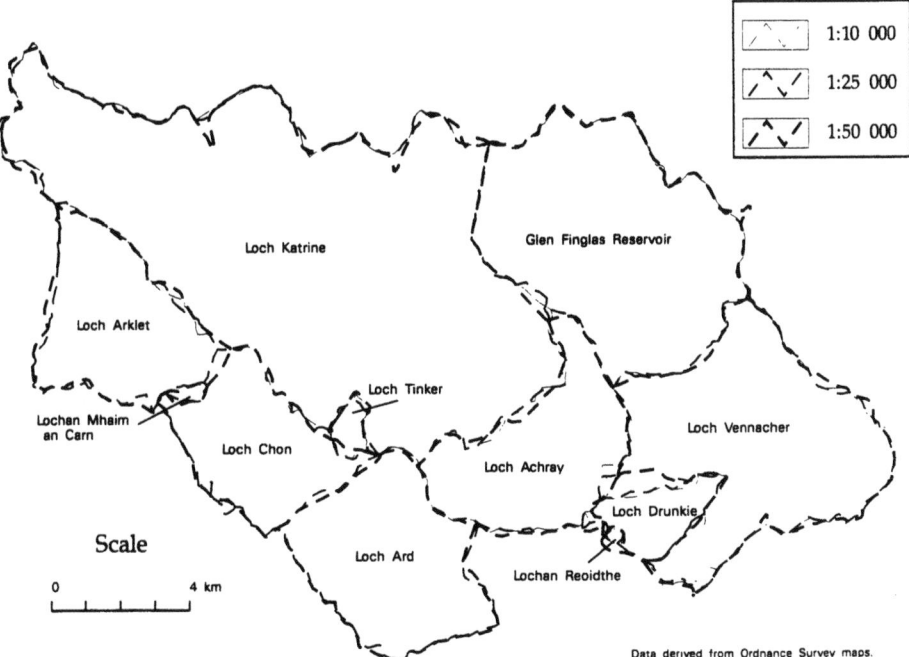

	1:10 000
	1:25 000
	1:50 000

Loch Katrine

Glen Finglas Reservoir

Loch Arklet

Loch Tinker

Lochan Mhaim
an Carn

Loch Chon

Loch Vennacher

Loch Achray

Loch Drunkie

Scale

Loch Ard

Lochan Reoidthe

0   4 km

Data derived from Ordnance Survey maps.

Figure 5.1   Manual interpretations of catchment boundaries.

Several issues are associated with these observations:

1. Catchment boundary interpretation using topographic maps depends upon the representation of both altitude and water features. The level of detail of the topographic map is dependent upon the scale of mapping (both source and published scales) and the compilation guidelines used by the mapping organization. Since the compilation of the 1 : 25 000 map is based upon the 1 : 10 000 map, differences in details represented on these maps is due to cartographic generalization. The 1 : 50 000 map is derived as a separate product and thus the photogrammetric mapping and photographic interpretation are independent of that for the larger scale maps. The principle should be that the source map accuracy should be known as it is a reliable guide to overall accuracy.

2. Uncertainty in the compilation of altitude, hydrological or cultural details on the source map will influence the quality of the interpretation of a catchment boundary. One example of this is the difference between interpreted catchment boundaries of Loch Drunkie, due to the smaller vertical interval of the contour data at the larger map scale, allowing much greater accuracy in position in X–Y. Thus, the principle to follow is that the reliability of the boundary delimitation ultimately depends upon the integrity of the mapped data, either in graphical or digital form.

3. In areas of commercial forestry, on shallow slopes, the direction of drainage may be changed: for example, between Loch Achray and Lochan Reoidhte where forestry is on the catchment boundary. Thus, the validity of the boundary interpretation will vary according to the land cover and land use.

4. At medium scales (1 : 50 000) the level of generalization of the contours can have a significant impact on the reliability with which a boundary may be delimited.

At larger scales (1 : 25 000) the addition of boundary details such as walls can assist in the delimitation particularly these land parcel features which are *de jura* ownership boundaries and follow catchment boundaries. At larger scales (1 : 10 000) the level of contextual detail used in manual interpretation can be lower than at smaller scales. The consequence is greater scope for gross errors in boundary delimitation combined with greater accuracy in the detail of the delimited boundary. This is obviously a complex condition for which to determine overall accuracy.

5.3 CATCHMENT DATABASE

A database of lochs in Scotland has been compiled with their associated catchments. The principles above are incorporated into the dataset definition. The database contains chemical analysis of the loch waters and physical characteristics of the water catchments. In addition, there are several fields in which records of sampling dates and organization responsible and gazetteer information are held.

Most standing waters greater than 1 km^2 in surface area are included (a total of 170) plus an additional 400 lochs distributed across Scotland. The source of the lochs was the Land Cover of Scotland 1988 (LCS88) digital dataset (Macaulay Land Use Research Institute, 1993), part of which is based upon Ordnance Survey 1 : 25 000 scale maps. The LCS88 dataset includes all standing waters greater than 2 ha in area: a total of 7600, of which approximately 2.5 per cent are greater than 1 km^2. The positional accuracy associated with the boundaries of lochs is approximately ± 12.5 m (Macaulay Land Use Research Institute, 1993).

The land cover within each catchment has been assessed using the LCS88 (Macaulay Land Use Research Institute, 1993; Aspinall and Pearson, 1993). Trees can be described in approximately eight levels of density (plus the presence of clumps or lines of trees). Thus, the areal extent of tree covered land can be assessed for each catchment and comparisons made between the different catchments in terms of the nature of the tree cover. Plate 1 (see colour plates section) shows the percentage of land on which trees are present (at a density of greater than two stems per ha over an area of 10 ha), excluding clumps of trees below 2 ha in area.

A national validation exercise for the LCS88 dataset produced comparisons between the land cover interpreted from the photograph and that recorded during field visits (Macaulay Land Use Research Institute, 1993; Aspinall and Pearson, 1993). The percentage of correctly interpreted land cover classes at the main class level of the key was weighted by the area of each class within each catchment (Veregin, 1989). The output was a table of the percentage of correctly interpreted land cover within each catchment (at that level in the classification scheme).

The calculation of land cover reliability can be undertaken for different levels of the classification hierarchy. Therefore, if a particular land cover feature is of interest, such as the interpretation of trees, the error can be assessed for all classes with trees present; catchments can be described according to the extent and accuracy of interpretation.

The contents of Table 5.2 point to the consequences of including some measures of uncertainty within studies of catchments. For catchment area, Loch Katrine was attributed as having the highest reliability of estimate, whereas Loch Drunkie had the lowest. In contrast, from the estimates of reliability of land cover interpretation

Table 5.2 Example of ranking of catchments according to reliability of environmental characteristics

Name	Area	Land cover	Rainfall
Loch Katrine	1	4	5
Loch Vennachar	2	5	2
Loch Finglas	3	2	4
Loch Achray	4	3	3
Loch Drunkie	5	1	1

1 = highest ranking (most accurate), 5 = lowest ranking (least accurate).

within each catchment, the highest reliability was associated with Loch Drunkie. This is because coniferous woodland occupies all the land within the Loch Drunkie catchment and coniferous woodland is accurately interpreted in LCS88.

5.3.1 Rainfall volumes

Assessment of net catchment water inputs, storage and loss is derived from models using rainfall and evapo-transpiration. Data for these are derived by geo-statistical techniques (Matthews *et al.*, 1994; Hudson and Wackernagel, 1994), and river flow and reservoir data are also used. The rainfall model uses data for 30-year monthly averages (1941–1970) from 1500 rainfall stations across Scotland. The methodology employed is a hybrid of a stepwise multiple linear regression and point kriging of the residual values. A surface representing uncertainty in the levels of rainfall in the model has been generated at a resolution of 1 km^2 as a product of the kriging methodology, providing a means of expressing a level of confidence with estimates of rainfall in each 1 km^2 (Aspinall and Miller, 1990; Matthews *et al.*, 1994) (Plate 1). The model presents the rainfall in millimetres per annum, assumed to be equal in value evenly across the cell, for a ground surface area estimated as a point.

A ranking of the estimates of input rainfall volumes for five catchments is shown Table 5.2. Loch Drunkie has the highest level of reliability and Loch Katrine the lowest. This is, principally, a consequence of the greater range in altitude within the Loch Katrine catchment and thus a greater cumulative uncertainty in the rainfall estimates, and the greater distance from the nearest gauge used in the model of rainfall, at Strathyre (10 km North-East of the catchments listed).

The catchments are used as areal units for which the average annual volume of rainfall are estimated by intersecting the rainfall and catchment datasets and calculating the product of the proportion of each cell within a catchment and the rainfall levels. The output of this operation is a lumped estimate of rainfall volume for the entire catchment.

This procedure is followed by three further operations:

1. estimate of uncertainty in rainfall for each cell
2. proportions calculated for the minimum and maximum estimated area of each catchment
3. combining rainfall and area uncertainty for the accuracy for the input for the catchment.

Several limitations exist with respect to the use of the rainfall data:

1. resolution: there will be variation in rainfall levels within each 1 km^2 cell
2. distribution of rainfall stations: measurements of rainfall distant from stations are likely to be less accurate than measurements at stations
3. local variation (over areas greater than 1 km^2: the model is not sensitive to local variations in altitude, nor the sheltering effects of topography or aspect
4. time period of the rainfall values used: comparisons between the rainfall model and outputs to catchments are for annual mean input values between 1941 and 1970.

No allowance was made for interception and precipitation of water vapour in mist or for interception storage on trees and their evapo-transpiration, which are likely to be less significant in volumetric terms but may be significant with respect to transportation of input pollutants (Ferrier *et al.*, 1990). Further work is required in the estimation of the volume of water which is re-directed from one catchment to another by the Hydro-Electric Company. In certain catchments this can be a significant volume of water (P. Donaldson, personal communication) and typical seasonal volumes can be built into the database as links between catchments.

5.3.2 Soils

The soils in this area are mapped at a scale of 1 : 50000 and due to the complexity of the geology and topography, the soil map units generally comprise more than one soil type. While recent work (Boorman *et al.*, 1994) has led to an assessment of the proportions of each soil type within each map unit, this has been largely based on expert judgement and, as yet, remains invalidated. It is unlikely that these proportions will remain constant throughout: e.g. it is known that the proportion of peat soils in some map units declines as the slope increases. It is also likely that each map unit will include some minor impurities: i.e. other soil types of limited extent.

The allocation of soil attributes to each of the soil types within the area was based on a deriving a modal or representative soil profile from a Scotland-wide database of over 12000 soil profile descriptions (and over 9000 profiles with accompanying chemical analyses). However, by deriving a modal value in this way, it is clear that neither the regional variation nor the variation inherent in any classification system is taken into account, and the assessment of soil attributes for each soil in the catchment remains deterministic.

As the soils in a catchment have a major impact on both river flows and stream chemistry, a hydrological model, HOST (Boorman *et al.*, 1994), has been used to apportion incoming precipitation to surface runoff (no likelihood of chemical amelioration) and to infiltration (some chance of hydrological and chemical buffering). This model has 29 classes and as it was developed from the representative profiles it can be directly linked to the distribution of soil types within any catchment by simply recoding the soil map units. Each of these classes has been allocated a value for the hydrological index Standard Percentage Runoff (SPR). This parameter is used by hydrologists to determine the proportion of precipitation in any one event that will directly runoff and contribute to flood flow; however, it can also be used to determine which proportion of the precipitation infiltrates into the soil and therefore has the chance of altering the chemistry of the soil solution.

5.4 DISCUSSION

The outcome of this study has been an assessment of the reliability of the results for those catchments for which rainfall, soils and land cover information are used for modelling water quality. The issues associated with delimiting catchments point to further research into automated techniques that make use of more information than just digital elevation models (Beven and Wood, 1994). One option may be to employ a rule-based approach to incorporating the contextual information offered by the topographic map in association with the elevation data. Such a rule base may be embedded within an expert system. However, such an approach on its own remains vulnerable to the underlying inaccuracy of the map data.

At small scales, over large geographic areas, errors in delimitation of large water catchments are likely to be insignificant, however, at large scales, small catchments or sub-catchments of larger catchments will be subject to greater relative error. Although the examples provided in this paper illustrate some of the implications of such errors, the levels of uncertainty of other factors associated with the catchment may be much more significant.

Current work is assessing the reliability of water catchment delimitation using automated techniques. This is being carried out using Ordnance Survey 1 : 50 000 elevation data for a range of topographic conditions. The objectives are to produce estimates of reliability of the estimation of physical factors within each catchment in the database. This will provide a look-up table against which models that use catchment, land cover, soils and mapped river data can be matched, and will as a guide assessments of the reliability of such modelling.

ACKNOWLEDGEMENTS

The authors would like to acknowledge the assistance of Keith Matthews, Gordon Hudson and David Elston for their work on the rainfall data for Scotland and Richard Birnie for comments on drafts of the paper. Data on water volumes for rivers or catchments with reservoirs were supplied by the North East Purification Board and Hydro-Electric Plc. Funding of the project was from the Scottish Office Agriculture and Fisheries Department.

REFERENCES

ASPINALL, R. J. (1994) GIS and spatial analysis for ecological modelling. In *Environmental Information Management and Analysis: Ecosystem to Global Scales*, ed. MICHENER, W., BRUNT, J. W. and STAFFORD, S. G. Taylor & Francis, pp. 377–398.

ASPINALL, R. J. and MILLER, D. R. (1990) Mixing climate change models with remotely-sensed data using raster based GIS. In *Remote Sensing and Global Change*. Proceedings of the 16th Annual Conference of the Remote Sensing Society, Swansea, 1990.

ASPINALL, R. J. and PEARSON, D. (1993) Data quality and spatial analysis: analytical use of GIS for ecological modelling, In *Proceedings of GIS and Environmental Modelling*, NCGIA Meeting, Breckenridge, September 1993.

BAND, L. E. (1989) A terrain-based watershed information system. *Hydrol. Processes* **3**, 151–162.

BEVEN, K. J. and WOOD, E. F. (1994) Catchment geomorphology and the dynamics of runoff contributing areas. *J. Hydrol.* **65**, 139–158.

BOORMAN, D. B., GANNON, B., GUSTARD, A., HOLLIS, J. M. and LILLY, A. (1994) The hydrological aspects of the HOST classification of soils. Institute of Hydrology report to MAFF.

FERRIER, R. C., JENKINS, A., MILLER, J. D. and WALKER, T. A. B. (1990) Assessment of wet deposition mechanisms in an upland Scottish catchment. *J. Hydrol.* **113**, 285–296.

HUDSON, G. and WACKERNAGEL, H. (1994) Mapping temperature using kriging with external drift: theory and an example from Scotland. *Internat. J. Climatol.* **14**, 77–91.

MACAULAY LAND USE RESEARCH INSTITUTE (1993) The Land Cover of Scotland by aerial photographic interpretation: Final Report. MLURI, Aberdeen, Scotland, 1993.

MAIDMENT, D. R. (1993) Developing a spatially distributed unit hydrograph by using GIS. *Internat. Assoc. Sci. Hydrol. Public.* **211**, 181–192.

MATTHEWS, K., MACDONALD, A., ASPINALL, R. J., HUDSON, G., LAW, A. N. R. and PATERSON, E. (1994) Climatic soil moisture deficit – climate and soil data integration in a GIS. *Climate Change* **28**, 273–287.

MOORE, I. D., GRAYSON, R. B. and LADSON, A. R. (1991) Digital terrain modelling: a review of hydrological, geomorphological and biological applications. *Hydrol Process.* **5**, 3–30.

MURRAY, J. and PULLAR, F. P. (1900) A bathymetric survey of the freshwater lochs of Scotland. *Scot. Geogr Mag.* **16**, 193–235.

ORDNANCE SURVEY (1992) *1 : 50 000 Scale Height Data User Manual.* Southampton: Ordnance Survey.

TOWERS, W. (1994) Towards a strategic approach to sewage sludge utilization agricultural land in Scotland. *J. Environ. Plan. Management* **37**, no 4. 447–460.

VEREGIN, H. (1989) A taxonomy of error in spatial databases. NCGIA Technical Paper 89–12, pp. 115.

PART TWO

Computational Support

CHAPTER SIX

Version management for GIS in a distributed environment

TIGRAN ANDJELIC and MICHAEL WORBOYS

6.1 INTRODUCTION

Geographical information systems (GIS) may be used to support a wide range of activities, from simple tasks such as inventory management to complex decision making and design activities. For example, a GIS used by a utility company may support not only simple queries regarding the location and condition of components of the supply network, but also complex decisions on the planning of new facilities. It is clear that the kind of technological underpinning required may be different for different task types. The focus of this chapter is the database support that is required for a GIS that is used for design activities. Such activities have special requirements, including the management of more than one alternative plan and the handling of transactions with the system that are much more complex and lengthy than traditional database interactions.

This chapter will discuss the requirements for GIS as a design tool and consider how well traditional GIS databases meet these requirements. Attention will be given to long transactions and version management. The key technical issue is the integration of database management regimes that may be quite different for different classes of applications. A traditional database is tuned to handle large numbers of concurrent but short transactions, such as in airline reservation systems. A design database, on the other hand, is constructed to manage small numbers of complex transactions. Next-generation systems should transparently support an integrated environment where both types of data management protocol have a place.

The chapter is structured to move from requirements to implementation. The next section will consider the requirements of a GIS that is to be used for design activities. Section 3 will provide background material on database consistency maintenance and version management. Sections 4 and 5 will discuss the concepts behind our approach to integration, and section 6 will briefly describe our implementation. The chapter concludes with some open problems and suggestions for further research.

6.2 GIS FOR DESIGN

From a data-management perspective, the majority of current GIS may be placed in two categories:

1. Enhancements of traditional databases with functionality for the handling of spatial data.
2. Evolutions from computer-aided design (CAD) systems. enhanced with general data-handling capabilities.

A primary characteristic of the first class is that its members support concurrent access with the traditional database approach to transaction management and do not support data duplication. Indeed the concept of database normalisation is the antithesis of data duplication. The second class emerged when CAD systems were also required to support the handling of large amounts of data. The main characteristic of this category is that its databases support data duplication. The need for functionality supported by the second class has been recognised and is well supported by some systems (Easterfield *et al.*, 1990). These GIS enable increased concurrency of data and access, both of which are among the prime requirements of data management systems for design applications.

As a first step in the identification of the database support needed for GIS in a design environment, the general requirements (Korth *et al.*, 1988) for a computer-based design environment are itemised:

- The division of a large design task into sub-tasks requires that the supporting database is correspondingly partitioned. In these cases, it is important to ensure shared access over the whole database and unrestricted read-only access to all partitions for privileged users and applications.
- Each partition is assigned to a group of designers. Data sharing and exchange requirements are very high while the conflict possibilities are limited, so the locking mechanism has to be quite relaxed. To accommodate the requirement for group work, some versioning mechanisms are highly desirable.
- Some of these sub-tasks are sub-contracted to external users, who will need some limited access to the database. Such users must be provided with sufficient meta-data in order that they may efficiently use the subset of the database that they require.
- There is a need for parallelism among designers' transactions so that they can be viewed as a transaction set rather than a serialised sequence. These transactions very rarely come into a conflict so checking for serialisability would impose unnecessary overheads. Once the transactions terminate it is possible to detect and resolve conflicts.

GIS add their own specific database requirements, to those given above. The first is the requirement for seamless mapping and hence for a logically seamless database that enables users to move freely over the whole database coverage without imposing any technical or implementation constraints. Secondly, GIS need to include additional databases so as to enrich their current data resources or to enlarge current data coverage by the size of the connected external databases. Thirdly there is a need for data versioning in order to enable the co-operative work of many designers at the same time.

6.3 CONSISTENCY MAINTENANCE AND VERSION MANAGEMENT FOR GIS

In order to maintain database consistency, each database item has to comply with a set of consistency constraints. Unless there are safeguards, consistency of the database can be violated through concurrent access to the same data item. For a standard database management system (DBMS) the problem of maintaining consistency is solved by enforcing concurrency control mechanisms. The serialisability criterion prescribes that any concurrent access (a set of transactions) may be serialised so that transactions performed by the concurrent users are executed one after the other in some order. A related concept is transaction atomicity that defines a transaction as a minimal operation performed on the database leaving it consistent. So, performing one transaction at a time will ensure that an originally consistent database will remain consistent. There are several serialisability-based concurrency mechanisms employed within traditional DBMS, as follows (Barghouti and Kaiser, 1991):

- *Two-phase locking:* a standard concurrency control mechanism (Eswaran *et al.*, 1976) which enforces serialisability by defining a locking mechanism for transactions. If the transaction requires access to a data item already locked by some other transaction, than it is forced to wait.
- *Timestamp ordering:* ensures serialisability of transactions by assigning a unique number to each transaction, called its timestamp. According to the values of timestamps, transactions are ordered and access to data granted.
- *Multi-version timestamp ordering:* a principle which allows a more relaxed concurrency control regime by allowing concurrent versions of data to exist and different transactions to lock different versions of the same data item so long as the transaction can see a consistent set of versions.
- *Optimistic non-locking:* another approach to relaxing concurrency control, where the locks for read-only access are not enforced and regarded as an unnecessary overhead for the system. If serialisability can be subsequently established through timestamps then locking was not necessary, but if the serialisability cannot be established then the database has to be rolled back to its previous correct state. This idea increases concurrency but for some cases it has serious disadvantages.
- *Nested transactions:* combines several transaction units into one transaction. This undermines the principle of atomicity of transactions allowing them to be composed of any number of simple atomic units or already composed transactions.

These approaches to concurrency control may be sufficient for traditional databases. However, they fail to deliver sufficient levels of concurrency control for databases regarded as non-traditional, such as databases used for CAD/CAM, CASE and GIS. For databases supporting such applications, issues relating to consistency and concurrency control mechanisms are quite different. Database consistency for these applications does not in general need to be as strict as with the traditional approach. The definition of database consistency may be relaxed so that it allows for a database to be consistent if each user at any time sees a consistent set of data versions. Of course, being able to see a consistent version of the database does not mean that each user has to see the same version, as would be the case with the traditional notion of consistency.

Research on this topic has identified a new set of requirements (Bancilhon *et al.*, 1985; Yeh *et al.*, 1987; Korth and Speegle, 1994) which could be related to GIS as follows:

- Support for *long transactions* is of a particular importance for design and GIS environments, where a transaction is extended in time, possibly lasting for days. It becomes very important to enable a concurrent execution of these long transactions without imposing a locking mechanism that would ensure serialisation.
- Support for *versioning of data*, so as to allow limited data sharing when data exchange among multiple users cannot be serialised. This enhances user control over long transactions and enables interactive changes to the transaction while it is not completed. A long transaction before it is committed is saved as a set of short transactions within an alternative (version) of the data.

In order to meet these requirements it is important to develop and implement new concurrency models and protocols. Many different models have been recently presented (Bancilhon *et al.*, 1985; Chou and Kim, 1986; Easterfield *et al.*, 1990; Katz, 1990; Barghouti and Kaiser, 1991; Korth and Speegle, 1994), including check-in/check-out, co-operating transactions, multi-level transaction models, and version control separated from concurrency control. In the last case, the principle employed is that when data is accessed for read-only, the system will return the most up-to-date copy, while when data is accessed for read–write, the system will make a new copy of the data. The version control mechanism ensures management of these data copies with regard to different users and that each user sees a consistent view of data versions at all times. Figure 6.1 describes one of the versioning mechanisms

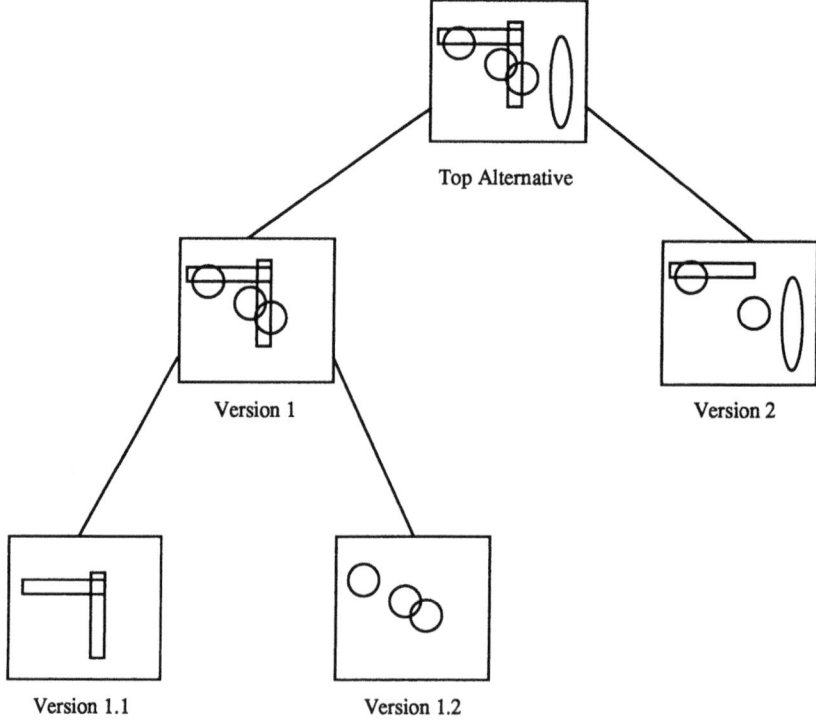

Figure 6.1 Versioning in GIS.

employed in Smallworld GIS in order to enable consistency of database views with a limited sharing of data. Each sub-alternative begins its life empty and dependent on its parent. The changes made in the alternative are private until posted to its parent. The changes made to the parent after the creation of the child are kept private within the parent until the child alternative merges them with its own data. In this way all alternatives are independent across the hierarchy and data propagation is left to the user.

6.4 MULTI-DATABASE SYSTEMS FOR GIS INTEGRATION

Figure 6.2 shows a schematic of a multi-database system with a GIS as a focal point. The GIS has access to external databases. The necessity of data sharing has been recognised in the GIS community (Laurini, 1995) and the results of research in distributed traditional databases (Sheth and Larson, 1990) are now being applied to GIS. There is a need to augment GIS information sources by means of externally connected databases. GIS enlarged in this way define a group of databases with a need for communication and data exchange. The group of connected databases via GIS is likely to form a heterogeneous and distributed environment. As described in Laurini (1995), several combinations of database connections could be found:

- Remote databases where each user has protocols to access remote databases on request.
- Distributed databases where all databases behave as a single virtual database with an agreed protocol for communication.

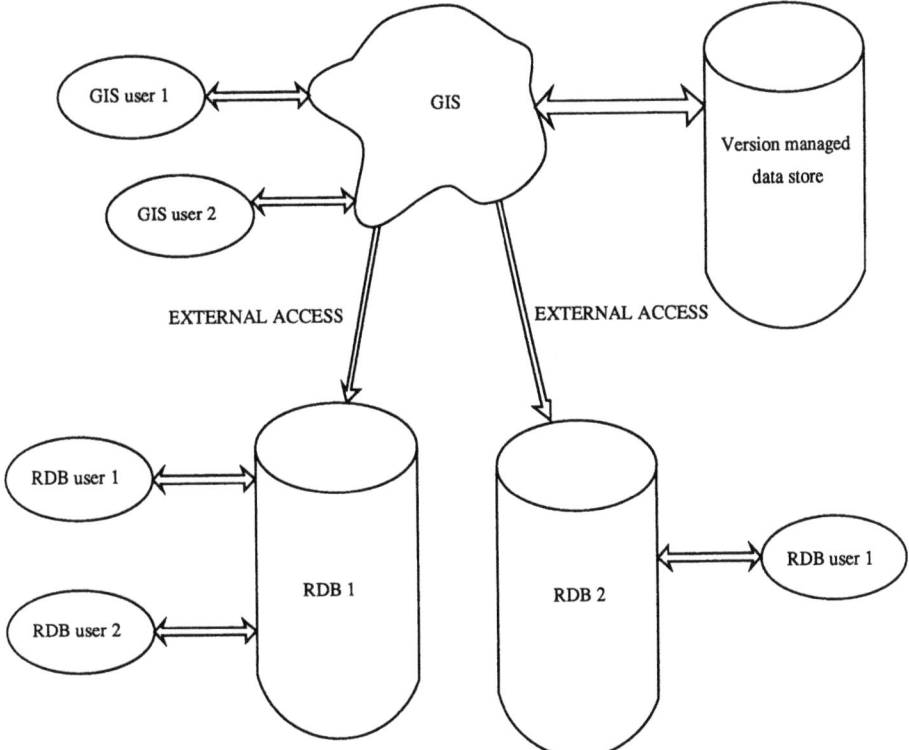

Figure 6.2 GIS with external database access.

- Homogeneous distributed databases with the same management system over the whole group of connected databases.
- Heterogeneous distributed databases where there are several management systems with an agreed protocol for communication.

The level of coupling among databases can vary and, as classified by Sheth and Larson (1990), it is possible to choose connections with several different degrees of participation. This chapter reveals a particular type of heterogeneity between databases with regard to the nature of transactions supported by them. We term this a *heterogeneity of transaction environments*.

As the taxonomy of heterogeneity differs among research papers, it deserves a short explanation. In order to consider a database set heterogeneous we have assumed that all databases could fall into one of two categories. The first category consists of DBMS that support version management and hence enable long transactions. The second category consists of databases in the traditional sense supporting short transactions in a single version environment. Any collection of DBMS with members in both categories is *heterogeneous*. This definition is important because the connection of heterogeneous databases imposes some fundamental constraints on GIS for design. The structural difference between these approaches to database management proves to be a barrier to extending GIS design functionality over shared data. In order to overcome this obstacle the chapter suggests an implemented working model of the heterogeneous environment with integrated transaction environments.

6.5 INTEGRATION OF TRANSACTION ENVIRONMENTS

Figure 6.3 shows the merge of the two-transaction environments that need to be supported for GIS design applications. The long transactions and short transactions are shown in separate bubbles that intersect in a shared environment containing data managed directly or indirectly by both database management systems. Two quite different database regimes based upon long and short transactions have been identified, and it is necessary to define an integration model that allocates a subset of data from each environment to a virtual environment such that an integration of

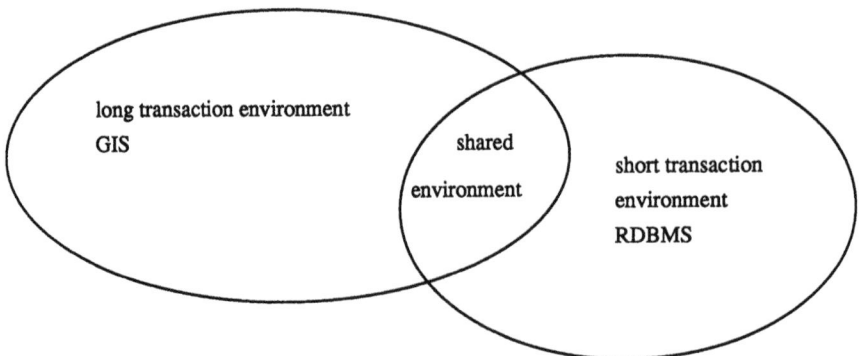

Figure 6.3 Merge of transaction environments in GIS.

functionality may be achieved. This integration is achieved by the *interface level*, defined as:

- A partial interface enabling one or both environments a read-only access to the other. This approach does not raise consistency problems and it does not integrate transaction environments. The only benefit is heterogeneous access of one database to another.
- A partial interface on a read–write basis implemented from one environment enabling it to extend into the other but not conversely.
- A full interface connecting the different environments on a read–write basis both ways, with defined behaviour.

If implemented as the third case, the interface will bring the possibility of short transaction management into the version-managed GIS database. So far, there have been no attempts to address these issues. As described in (Easterfield *et al.*, 1990) a GIS with version management, and thus support for long transactions, enhances the concurrency and design possibilities of GIS. However, the shortcoming of the model is that it supports only long transactions, which is the opposite from traditional DBMS. The importance of short transactions for any design environment has been neglected and there is no research that integrates these two different models. In order to tackle this issue within the heterogeneous environment a suitable model of change propagation must be developed. The model should describe scope and constraints for change propagation among versions.

Within this chapter we suggest a model that corresponds to the read–write partial interface enabling extension of the long transaction environment into the short transaction environment. It is possible to manage the shared data equally from either environment. The short transaction environment manages the master data while the long transaction environment manages the most recently updated copy. The model defines two levels of change propagation and hence two levels of conflict resolution. The first level performs completely within the version managed environment where the changes are posted to higher alternatives or merged down to lower level alternatives and all without change propagation to the external database. The second level represents change propagation into the external database. It takes effect when the version managed data-store becomes consistent across the version hierarchy. At that point, the top alternative is merged with the external database making the changes visible across the whole system. When the user chooses, at the final stage, to commit all changes in the top alternative, the second level ensures synchronisation of all data versions, both internal and external. The model is implemented by taking advantage of the facilities in both transaction environments and building a suitable protocol into the interface between them.

6.6   IMPLEMENTATION OF THE INTEGRATION MODEL

Figure 6.4 shows schematically our model of a heterogeneous distributed GIS, where the version-managed GIS accesses external relational databases with short transaction environments and a shared transaction environment between the version managed data-store and the relational database. The main idea is that there is a transaction environment shared between the version-managed GIS database and the short-transaction-managed external relational database.

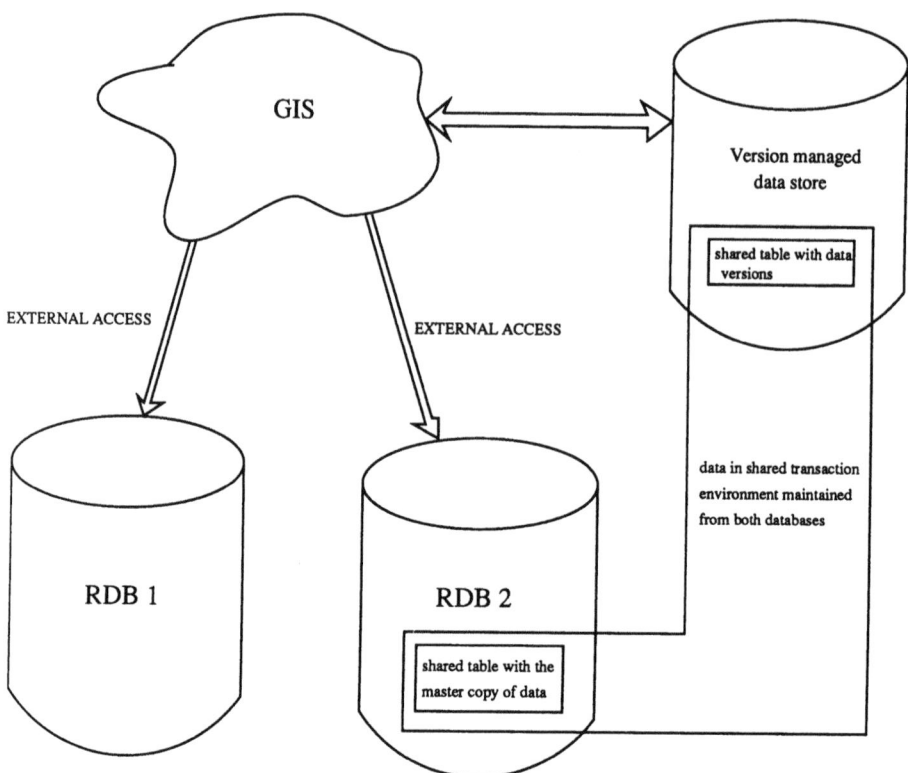

Figure 6.4 A schematic of a heterogeneous GIS with shared transaction environment.

The implementation is based on a version-managed environment provided by Smallworld GIS, with an external link to the relational DBMS Oracle. Before enhancement, the interface between the GIS and the external database is such that the external database is accessed from the GIS in read or write mode and the connection is treated as an ordinary user access. This basic interface enables GIS to perform short transactions with external databases. The shortcoming is that all changes become immediately visible across the version hierarchy and usefulness of those data for design is very limited. Our enhancement of this basic interface enables versioning of data kept in the relational database using the GIS version control. In order to achieve this it is important to change the behaviour of the basic external interface so that it handles read/write accesses differently. The *copy-on-update* principle used is one of the basic concepts used within our GIS version-managed datastore. Whenever a read access is issued, the version controller retrieves the most up-to-date version of the data accessible by that user, and whenever a write access is issued the version controller makes a copy of the record, updates it and amends all pointers to it. The newly created record becomes a part of the user's alternative as a private copy not visible from other alternatives. If the private copy is updated, then a new version of that copy is not created. When alternatives are merged and conflicts are resolved, the data copy that is most up-to-date becomes visible throughout the system as the master copy. This principle, if extended to external connections, could provide version management of relational databases. Since it is not straightforward to implement this using relational database languages, it is necessary to store copies of updated records within the GIS data-store and master copies in the external database.

As it is implemented in our working example, the bulk of data is kept in the Oracle tables while the updated versions are stored in shadow tables within the Smallworld data-store. The shadow tables are made to be structurally compatible with their corresponding Oracle tables. Two physical parts of the virtual table are maintained transparently with the enhanced interface behaving as one virtual table. Each time an access is issued, the interface would reference both physical parts and deal with records according to the copy-on-update principle. The Oracle master table is used for read-only purposes while the updates are placed in the Smallworld shadow table. Only when the status of the record changes from private to public (when posting to the top alternative) is the Oracle table updated and conflicts at that level resolved. The physical division of data between master records and versions divides the process of conflict resolution in two phases, as follows:

- The first phase merges alternatives within the Smallworld version-managed data-store (VMDS) with conflicts resolved in the standard way.
- The second phase posts data from the lower alternative to the top alternative in the GIS and maintains consistency with the Oracle master table. This phase is interactive, with the user deciding on the means to resolve conflicts.

For the short transaction environment only master records exist, while the long transaction environment is aware of the different copies from the data-store. The conflicts in the second phase are reduced to relational database conflicts but with scope for interaction with the user. A universal collection of criteria for conflict resolution is neither practical nor desirable. The precise form of conflict resolution will depend on the details of the specific application and should be left to the user.

6.7 CONCLUSION

The current state of the art provides only a minimal interface between GIS and external relational (or other) databases. This chapter has proposed a model that integrates more fully the long and short transaction environments provided by a GIS and a relational database, respectively. We also described an implementation using Smallworld GIS and Oracle RDBMS. The management of the schema and tables of each external relational database from within the GIS is a further research question. Also, work is required on a model of propagation of updates from the external database to the version managed data-store, and a GIS server to support such a model.

ACKNOWLEDGEMENTS

The authors gratefully acknowledge Smallworld and EA Technology for support during the course of this research. The chapter benefited from the comments of Chris Yearsley on a draft.

REFERENCES

BANCILHON, F., KIM, W. and KORTH, H. (1985) A model of CAD transactions. In Proceedings of the 11th VLDB Conference, Morgan Kaufmann, pp. 25–33.

BARGHOUTI, N. S. and KAISER, G. E. (1991) Concurrency control in advanced database applications. *ACM Comput. Surv.* **23**, 269–318.

CHOU, H. and KIM, W. (1986) A unifying framework for version control in a CAD environment. In Proceedings of the 12th VLDB Conference, Kyoto, Japan, pp. 336–344.

EASTERFIELD, M., NEWELL, R. and THERIAULT, D. (1990) Version management in GIS: Applications and techniques. Proceedings European Geographical Information Systems (EGIS) Annual Conference, EGIS Foundation, Utrecht, Netherlands.

ESWARAN, K., GRAY, J., LORIE, R. and TRAIGER, I. (1976) The notions of consistency and predicate locks in a database system. *Communications ACM* **19**, 624–632.

KATZ, R. H. (1990) Towards a unified framework for version modeling in engineering databases. *ACM Comput. Surv.* **22**, 375–408.

KORTH, H. F. and SPEEGLE, G. (1994) Formal aspects of concurrency control in long-duration transaction system using the NT/PV model. *ACM Transac. Database Syst.* **19**, 492–535.

KORTH, H. F., KIM, W. and BANCILHON, F. (1988) On long-duration CAD transactions. *Inform. Sci.* **46**, 73–107.

LAURINI, R. (1995) Distributed geographic databases: an overview. In GREEN, D. R. and RIX, D. (eds), *The AGI Source Book for Geographic Information Systems*. Association for Geographic Information, London, England, pp. 45–55.

SHETH, A. P. and LARSON, J. A. (1990) Federated database systems for managing distributed, heterogeneous and autonomous databases. *ACM Comput. Surv.* **22**, 183–236.

YEH, S., ELLIS, C., EGE, A. and KORTH, H. (1987) Performance analysis of two concurrency control schemas for design environments. Technical Report STP-036-87, MCC, Austin, Texas.

CHAPTER SEVEN

Programming spatial databases: a deductive object-oriented approach

NORMAN PATON, ALIA ABDELMOTY and HOWARD WILLIAMS

7.1 INTRODUCTION

Geographical Information Systems present significant challenges to database technology. The data which has to be store is structurally complex, and can only be described satisfactorily by expressive data models with explicit support for spatial data types. The analyses which are applied to geographic data are generally complex, requiring extensive access to derived relationships between spatially distributed concepts. Commercial relational databases have been found to be inadequate for supporting geographic applications because of their spartan data modelling facilities and the limited computational power of their query languages. This has lead to the use of coupled systems, where the spatial data manager and the database are distinct components, with consequent disadvantages for programmer productivity and run-time performance. However, while it is clear why the relational model is less than ideal for use with geographic data, the challenge remains of identifying a database technology which can provide, or be extended to provide, comprehensive and efficient spatial data management facilities.

This chapter addresses the provision of effective database support for spatial data management. Section 7.2 identifies a range of facilities which it would be desirable for a spatial database system to support, and evaluates a number of representative proposals against these requirements. Section 7.3 considers how such facilities should be catered for within database systems in order to obtain satisfactory performance, and compares the software architectures used by different proposals. Section 7.4 describes experience using a deductive object-oriented database system for managing spatial data, and section 7.5 draws some conclusions.

7.2 REQUIREMENTS – WHAT FUNCTIONALITY SHOULD A SPATIAL DATABASE POSSESS?

Proposals have been made which seek to provide database support for the geographic domain in a number of ways. For example, the relational data model can be extended with spatial data types and operations, as in GRAL (Guting, 1988) and GEO-SAL (Svensson and Zhexue, 1991); geographic data types and display facilities can be implemented using a database programming language, such as Napier (Kuo, 1994); geographic data management routines can be layered on top of an object-oriented database system, such as ONTOS (Roberts and Gahegan, 1993) or O2 (Scholl and Voisard, 1992); a deductive object-oriented database system, such as ROCK & ROLL can be used to describe and reason about spatial concepts (Abdelmoty et al., 1994). Such approaches can then be characterised according to criteria such as the following:

1. Aspatial data modelling facilities (ADM) – is the underlying aspatial data model appropriate for describing the structural characteristics of the domain?
2. Spatial data modelling facilities (SDM) – is support provided for a comprehensive collection of spatial data types and operations?
3. Declarative query facilities (QL) – can ad-hoc enquiries be made of a database without the need to write code in a programming language?
4. Programming language facilities (PL) – can arbitrarily complex analyses and computations be expressed using a language which is fully integrated with the database?

Table 7.1 illustrates how representative examples of different kinds of system, such as GRAL, Napier, ONTOS and ROCK & ROLL compare against these criteria. While caution has to be displayed when drawing conclusions from such a superficial analysis, the following points do emerge relating to the context within which a spatial database system may be developed:

- Extensions of the relational model with spatial data types do not overcome certain of its characteristic weaknesses, specifically the lack of effective data-modelling facilities and the absence of fully integrated programming mechanisms.
- Implementing spatial facilities using an imperative database programming language may be practical, but does not make geographic data widely accessible to non-programmers, and does not provide rich, high-level facilities for modelling the aspatial aspects of a geographic application.

The wide scope of geographic applications, which mix spatial and aspatial data types, ad-hoc querying and application development, is often inadequately catered for by existing database systems.

Table 7.1 can be seen as identifying certain necessary, if not sufficient characteristics for effective spatial data management. Many important issues are not addressed, for example the question of which spatial types and operations should be supported. The focus of this chapter is more on architectural considerations which

Table 7.1 Naive classification of SDB functionality

	GRAL	Napier	ONTOS	ROCK & ROLL
ADM			✓	✓
SDM	✓	✓	✓	✓
QL	✓			✓
PL		✓	✓	✓

should be taken into account when developing spatial database systems, as addressed in the following section.

7.3 ARCHITECTURE – HOW SHOULD SPATIAL DATABASES BE BUILT?

A weakness of the classification presented in Table 7.1 is that it gives no insight into the underlying software architecture. By contrast, Figure 7.1 illustrates where spatial types fit into the overall architecture of three representative spatial database systems. In this figure, QL indicates that a declarative query language can be used to access spatial data, PL indicates that a computationally complete programming language is supported for manipulating spatial concepts, and Kernel indicates the presence of spatial data types and operations at a low level within the database system.

In GRAL, the kernel of the database system has been extended with support for a range of spatial data types and operations that can be invoked from the algebraic query language. This means that spatial types are recognised and accounted for by important components of the database system, including the optimiser and the storage manager, and that efficient spatial analysis should result. The principal disadvantage is the fixed nature of the spatial facilities. As spatial types are supported within the kernel of the database, only the developers of the system can readily extend or change key parts of the implementation. Furthermore, the absence of any fully integrated programming mechanism means that it is likely to be extremely difficult for users to perform tasks that were not anticipated when the system was designed.

In ONTOS, a collection of spatial data types has been implemented using its application development facilities, which are essentially those of C++. The spatial constructs provided are far from fixed, and could easily be extended or revised to suit the needs of specific applications. However, there are minimal query language or optimisation facilities in ONTOS, and providing declarative access to the spatial data types would be a major task. The architecture of Napier would be depicted in the same way as that of ONTOS in Figure 7.1.

In ROCK & ROLL most of the spatial data handling facilities have been implemented in ROCK, an imperative database programming language, and thus can be readily extended, as in ONTOS. There is also declarative access to such data using

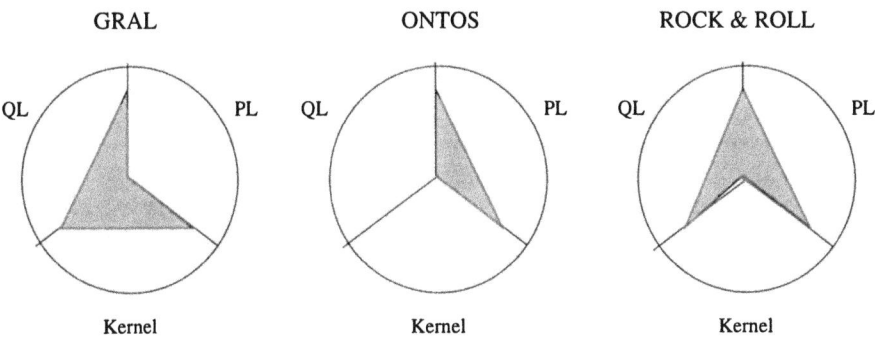

Figure 7.1 Context for spatial types in representative systems.

the logic query language ROLL. However, as the spatial types are supported using ROCK, rather than as part of the kernel of the system, the optimiser does not recognise the distinctive semantics of spatial operations, and thus is not able to perform certain of the transformations which could be supported if spatial data types were part of the kernel (Aref and Samet, 1991). Furthermore, this absence of kernel support for spatial concepts reduces the speed with which spatial operations are performed.

Two general points on architectures for spatial databases can be identified at this stage: (1) Kernel support for spatial data types is likely to yield improved performance, as the underlying database can organise spatial data in an appropriate manner, and because the query optimiser can take into account the semantics of such data. (2) Certain architectures provide a fixed set of spatial data types and operations that are difficult to extend. This is unfortunate, because different applications have different requirements, and it is difficult to anticipate all the uses to which a spatial database system may be put.

The following section presents an approach to spatial data management based upon the deductive object-oriented database system ROCK & ROLL, and shows how the architecture of this system can be revised to address the problems of existing spatial database systems presented in sections 7.2 and 7.3.

7.4 EXPERIENCE – A DEDUCTIVE OBJECT-ORIENTED DATABASE FOR SPATIAL DATA MANAGEMENT

This section presents our experience in applying the deductive object-oriented database (DOOD) system ROCK & ROLL to spatial data management.

7.4.1 Motivation

It is widely recognised that both deductive (Abdelmoty et al., 1993; Jones and Luo, 1994) and object-oriented (Worboys et al., 1990; Scholl and Voisard, 1992) facilities can be exploited to support effective utilisation of spatial data. There are, however, very few systems which seamlessly combine these two paradigms in a database context. The ROCK & ROLL system does, in a manner described more fully in (Barja et al., 1994, 1995), using the architecture presented in Figure 7.2.

The principal components of the system are an object-oriented data model (OM), an imperative database programming language (ROCK) and a logic language for writing inference rules and expressing queries (ROLL). The two languages can be seamlessly combined, because they share the same underlying data model, to yield what can be considered to be an integration rather than a coupling of the two programming paradigms. As neither language takes precedence over the other, characteristic weaknesses of single-language systems such as GRAL or ONTOS are overcome.

The motivation for this work is thus to establish the extent to which this particular blend of database functionalities is able to support effective development of complex geographic applications.

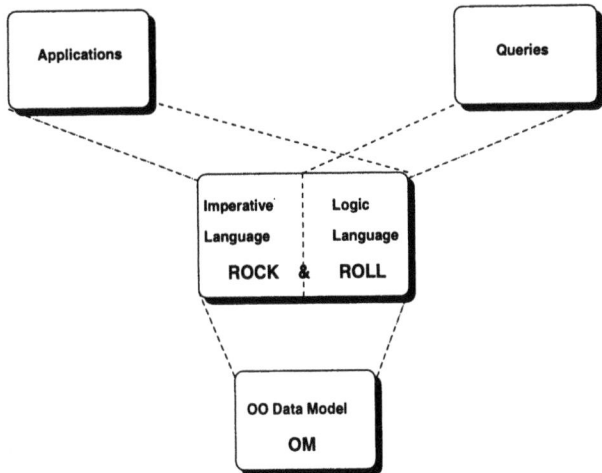

Figure 7.2 Relationship between the principal components in ROCK & ROLL.

7.4.2 Experience

The exploitation of ROCK & ROLL in the area of spatial data management has proceeded in two stages – the development of generic facilities for describing the spatial aspects of a domain, and the evaluation of these facilities using a resource allocation application. The remainder of this subsection outlines how ROCK & ROLL has been used to implement this application, with an emphasis on how the functionalities introduced in section 7.2, and used to classify different proposals in Table 7.1, are supported.

Structuring spatial data using OM

OM is the object-oriented data model of the ROCK & ROLL system. As such, it provides a range of mechanisms for describing the structural aspects of a domain, such as a series of type constructors (for sets, sequences and aggregates), and mechanisms for supporting the sharing of information through inheritance. As OM is the sole vehicle for describing structure in ROCK & ROLL, it is used for organising both spatial and aspatial information.

To give a flavour of how can be applied, Figure 7.3 gives examples of the structures of object types associated with the resource allocation problem. A number of modelling constructs from OM are illustrated in the figure. For example: *Structured Types* (drawn using rectangles) are used to represent any structured domain concept, such as roads or population_centres; *Primitive Types* (drawn using ellipses) are used to represent any scalar valued domain such as a price (which is stored as a real number); *Attributes* (drawn using circles) are used to represent the stored properties of a type, such as the price and availability of a land_parcel; *Sequences* (drawn using double circles) are used to build new types as ordered collections of items from existing types, such as path, which is represented as a sequence of flow-segment objects; *Specialisation* (drawn using a triangle inside a circle) which is used to indicate that one structured type is a subtype of another, and thus should inherit its properties, such as motorway, which is a subtype of road.

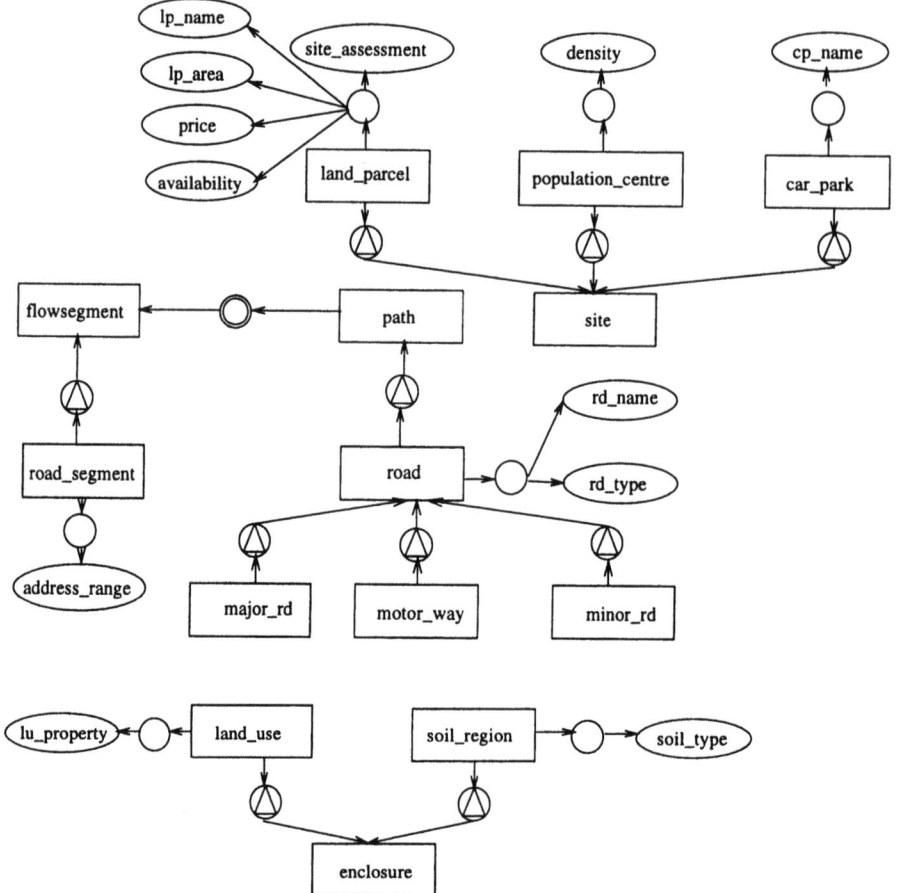

Figure 7.3 Example object types on the feature level of representation.

It can be seen from Figure 7.3 that OM provides data modelling facilities which are capable of describing in a direct way the structural characteristics of a representative collection of spatial concepts. In practice, OM has been used to structure both primitive (point, line, polygon) and application-specific concepts (carpark, supermarket) using the same data modelling facilities. That object-oriented modelling facilities can be applied to the representation of spatial data is not new, but few other implemented object models provide such a direct representation of semantic data modelling facilities as OM.

Defining operations using ROCK

ROCK is an imperative object-oriented database programming language, which can be used to create and manipulate objects structured using OM. The facilities supported by ROCK are comparable to those of other object-oriented database programming languages, with constructs for assignment, iteration (while, foreach), condition testing (if then else), input/output, etc. As such, ROCK can be used to implement geometric operations over spatial types, to load data into the database, and to build complete applications which exploit lower level spatial concepts. ROCK operations can be defined which are attached to OM types as methods; such operations are then stored in the database alongside the data to which they relate.

As an example of a ROCK operation at the geometric level, the method point_dist_point is attached to the type point, and returns the distance between the point to which it is attached and the point given as a parameter:

type point
```
  ...
  point_dist_point(pt: point): bool
  begin
    var xdiff := get_xCoordinate@pt - get_yCoordinate@self;
    var ydiff := get_yCoordinate@pt - get_yCoordinate@self;
    sqrt((xdiff * xdiff) + (ydiff * ydiff))
  end
end-type
```

In the above code fragment, the operator @ is used to invoke an operation. Thus get_xCoordinate@pt returns the value of the xCoordinate of the point pt. The result of the operation is the result of the expression given in the last line, i.e. sqrt((xdiff * xdiff) + (ydiff * ydiff)). This operation can then be used from programs written in either ROCK or ROLL.

The aim of this short description is not to give an introduction to how complete programs or spatial primitives can be written in ROCK, but rather to give a flavour of the facilities provided. Essentially, as a computationally complete language, ROCK can be used to carry out any programming task. This, plus the full integration of ROCK with OM means that complete applications can be developed without resorting to multi-component development environments, and that the facilities provided for manipulating spatial types can be readily extended using the same language as is used to develop user applications. Furthermore, the close association between operations and the data to which they relate that is encouraged by the object-oriented paradigm allows both to be stored together in the database for sharing among different applications and users.

Defining operations using ROLL

ROLL is a deductive database language (Ceri *et al.*, 1990), which supports logical inference of new facts from those stored in an OM database. The facilities offered by ROLL are similar to those offered by other deductive databases (Ceri *et al.*, 1990) except that ROLL is associated with an object-oriented data model, and is integrated with an imperative programming language, namely ROCK. As such, ROLL can be used to express queries over OM databases, and to describe derived properties of database concepts using a declarative rule-based style.

A number of authors have proposed the use of deductive capabilities for reasoning about spatial data (Abdelmoty *et al.*, 1993; Jones and Luo, 1994), and it has been suggested that such facilities are suitable for supporting spatial reasoning, defining derived properties of objects, rule-based feature recognition, and the expression of ad-hoc queries about spatial and aspatial aspects of the data stored in a database.

Spatial reasoning has been proposed recently as a complementary mechanism to computational geometry for the automatic derivation of spatial relations which are not explicitly stored (Egenhofer, 1991; Cui *et al.*, 1993). As an example of the use of ROLL, the following rules derive an orientation relationship between two objects of type site:

```
type site
   ...
   east(site)
   begin
      east(St2)@St1:-St2 isin get_east_site_set@St1;
      east(St3)@St1:-east(St2)@St1, east(St3)@St2;
   end
end-type
```

The first clause states that a site object St2 is east of another St1 if it is a member of the set of sites obtained by evaluating the get_east_site_set operation of St1. This latter operation could either retrieve values stored as a property of site, or could compute the members of the set using a geometric algorithm written in ROCK. The second clause states that St3 is east of St1 if St3 is east of another site object St2 and St2 is east of St1, thereby computing, by logical inference, the transitive closure of the east relationship.

As an example of a ROLL query, rather than a method, the following statement assigns to cs the set of carparks which are within 1000 metres of a population centre with a stored density greater than or equal to 2.0:

```
var cs := [{Carpark} |
      get_density@Pop = = D, D > = 2.0, get_position@Pop = = Pt1,
      get_position@Carpark = = Pt2, pt_dist_pt(Pt2)@Pt1 < = 1000.0];
```

The result of this query is the set of bindings obtained for the logic variable Carpark when the query after the | is evaluated. The query uses a number of features: attribute values are retrieved and unified with variables (e.g. get_density@Pop = = D); constraints are imposed on the values associated with a variable (e.g. D > = 2.0); and a method is invoked to verify that a distance is below a certain threshold (pt_dist_pt(Pt2)@Pt1 < = 1000.0).

The aim of this short description is not to give an introduction to writing queries and operations in ROLL, but rather to give a flavour of the rule-based programming style it supports. There is nothing that can be done in ROLL that cannot be done in ROCK, but the way in which retrieval tasks are described is considerably different, and the aim has been to provide two complementary mechanisms which can be used together to support complex tasks. A particular characteristic of ROLL is that its declarative nature means that queries and rules can be optimised by the system prior to evaluation, thereby relieving the user of the burden of planning efficient ways of carrying out a task.

7.4.3   Evaluation and plans

ROCK & ROLL can be reviewed against the criteria introduced in section 7.2 thus: aspatial data modelling (ADM) facilities are provided by OM, which includes a rich collection of object-oriented structuring mechanisms; spatial data modelling (SDM)

ROCK & ROLL

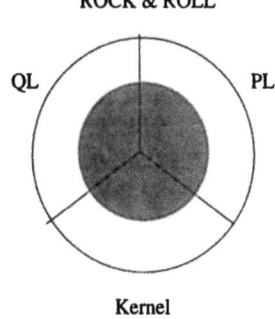

Kernel

Figure 7.4   Revising the architecture of ROCK & ROLL.

is supported by structures described using OM and operations implemented in a combination of ROCK and ROLL; query language (QL) functionality is available in ROLL, which also supports rule-based inference of derived properties; and programming language (PL) capabilities are available through the combined use of ROCK and ROLL, which can be used together in the development of complete data-intensive applications (Barja *et al.*, 1994).

The essential problems with the current approach to supporting spatial data using ROCK & ROLL relate to performance, and stem from: lack of spatial indexing in the current implementation of primitive types; lack of kernel facilities for efficient storage and retrieval of spatial types; lack of support within the optimiser for transformation of primitive spatial operations.

To address these problems, future work will revise the architecture of ROCK & ROLL to support spatial data types in the kernel of the system, as illustrated in Figure 7.4. This will yield the performance improvements for core spatial types which can be found in systems such as GRAL or GEO-SAL, but without constraining the extensibility of the system. The close association of the database programming facilities of ROCK & ROLL with the kernel-supported spatial types means that extensions to these built-in facilities can be achieved without resorting to the use of separate tools or subsystems.

7.5   CONCLUSIONS

This chapter has identified problems relating to the functionalities and architectures of existing spatial databases, showing that many systems emphasise one aspect of data management at the expense of others. A novel approach to spatial data management based upon a deductive object-oriented database seeks to avoid premature commitment to a limited set of facilities, without sacrificing performance through the use of purely application independent tools. Experience has shown the potential of the combined use of deductive and of object-oriented techniques, and ongoing work is addressing architectural issues which should yield competitive performance results.

ACKNOWLEDGEMENTS

We are pleased to acknowledge the support of the EPSRC AIKMS initiative (grant GR/J99360), and of Neil Smith and Glen Hart of Ordnance Survey as the Industrial

Partners. The following colleagues have worked on the development of ROCK & ROLL: Marisa Barja, Andrew Dinn and Alvaro Fernandes.

System availability. The ROCK & ROLL system is freely available over ftp. For details, see WWW page http://www.cee.hw.ac.uk/Databases/dood.html.

REFERENCES

ABDELMOTY, A. I., WILLIAMS, M. H. and PATON, N. W. (1993) Deduction and deductive databases for geographic data handling. In ABEL, D. and OOI, B. C. eds, *Advances in Spatial Databases (SSD '93)*, LNCS 692, 443–464. Berlin: Springer-Verlag.

ABDELMOTY, A. I., PATON, N. W., WILLIAMS, M. H., FERNANDES, A. A. A., BARJA, M. L. and DINN, A. (1994) Geographic data handling in a deductive object-oriented database. In KARAGIANNIS, D. ed. Proceedings of the 5th DEXA, 445–454. Berlin: Springer-Verlag.

AREF, W. G. and SAMET, H. (1991) Optimisation strategies for spatial query processing, In LOHMAN, G., SERNADIS, A. and CAMPS, R. eds, Proceedings of the 17th VLDB, 81–90, Morgan-Kaufmann.

BARJA, M. L., PATON, N. W., FERNANDES, A. A. A., WILLIAMS, M. H. and DINN, A. (1994) An effective deductive object-oriented database through language integration. In BOCCA, J. *et al.* eds, Proceedings of the 20th VLDB, 463–474. Morgan-Kaufmann.

BARJA, M. L., FERNANDES, A. A. A., PATON, N. W., WILLIAMS, M. H., DINN, A. and ABDELMOTY, A. I. (1995) Design and implementation of ROCK & ROLL: a deductive object-oriented database system. *Inform Syst.* **20**, 185–211.

CERI, S., GOTTLOB, G. and TANCA, L. (1990) *Logic Programming and Databases*, Berlin: Springer-Verlag.

CUI, Z., COHN, A. G. and RANDELL, D. A. (1993) Qualitative and topological relationships in spatial databases. In *Design and Implementation of Large Spatial Databases*, Third Symposium SSD '93, LNCS 692, 396–315. Berlin: Springer Verlag.

EGENHOFER, M. J. (1991) Reasoning about binary topological relations. In GUNTHER, O. and SCHECK, H. J. eds, *Advances in Spatial Databases*, 2nd Symposium, SSD'91, Lecture Notes in Computer Science, 525, 143–161. Berlin: Springer-Verlag.

GUTING, R. H. (1988) Geo-relational algebra: a model and query language for geometric database systems. In SCHMIDT, J. W. and MISSIKOFF, M. eds, *Advances in Database Technology- EDBT'88*, Lecture Notes in Computer Science, 506–527.

JONES, C. B. and LUO, L. Q. (1994) Hierarchies and Objects in a Deductive Spatial Database. In Proceedings of SDH, 588–603. London: Taylor & Francis.

KUO, Y.-J. (1994) Building a geographic information system in Napier88. Technical report, University of Glasgow, FIDE-2 BRA Project 6309.

ROBERTS, S. A. and GAHEGAN, M. N. (1993) An object-oriented geographic information system. Shell, Information and Software Technology, vol 35, no 10.

SCHOLL, M. and VOISARD, A. (1992) Geographic applications: an experience with O2. In *Building an Object-Oriented Database System: The story of O2, 585–618.* Morgan-Kaugmann.

SVENSSON, P. and ZHEXUE, H. (1991) Geo-SAL: a query language for spatial data analysis. In GUNTHER, O. *et al.* eds, *Advances in Spatial (SSD'91)*, 119–142. Berlin: Springer-Verlag.

WORBOYS, M. F., HEARNSHAW, H. M. and MAGUIRE, D. J. (1990) Object-oriented data modelling for spatial databases, *Int. J. GIS.* **4**, 369–383.

CHAPTER EIGHT

Spatial clustering with a genetic algorithm

MIKE HOBBS

8.1 INTRODUCTION

Many applications in geographical analysis require similar items to be grouped together according to some measure of their type and location. Once clustered, these items can be referred to by a label that is defined to embody the concept used to differentiate between the types. If the label is expressed as a rule, relating the properties of an item to its class, it can be used to classify further examples. Geographical Information Systems (GIS) are often used to form spatial groups by aggregating small areal units into larger, contiguous areas that can be given a particular classification. The well known Modifiable Areal Unit Problem (Openshaw, 1984) is present where aggregation at different scales produces different results from the same data. In common with other approaches, by Openshaw and others (Openshaw, 1991, Fotheringham and Rogerson, 1993), the clustering algorithm presented in this chapter reduces this problem by incorporating the areal unit in the model building process.

Genetic Algorithms (GAs) provide an approach that has no pre-determined bias towards particular sizes or shapes of clusters. They have been used to solve a wide range of problem types, including the induction of classifications from examples (Janikow, 1993), and are even beginning to appear in commercial applications (Kiernan, 1994). Their main attraction is the ability to apply a simple search technique, modelled on the biological processes of evolution and natural selection, to problems with vast numbers of potential solutions.

8.1.1 The spatial problem

The motivation for this research came from the desire to improve the efficiency and accuracy of solutions provided by another GA which was used to investigate the effect of location on residential property prices. This work by Cooley *et al.* (Cooley *et al.*, 1993) investigated a number of techniques to assess to what extent the price of a house was influenced by its location, in addition to its other characteristics such as size, condition, and the date of sale. They looked at how the various ways of accounting for location affected the error term in a model based on a regression equation of the house characteristics. The GA was designed to define areas of a city that had a significant effect on the price of a standardised unit of housing. The area classifications suggested by the GA reduced the error of the regression model and gave results of a similar quality to areas defined by an estate agent. However, it was thought that the GA could be improved by incorporating more heuristics that are generally applicable to problems of spatial grouping.

The original GA worked directly with real house sales data gathered from the City of Canterbury but required considerable processing for each potential solution suggested. A simplified set of test data was constructed to mimic the more interesting aspects of the spatial problem by pre-classifying houses into different types. This meant that area classifications suggested by the GA could be checked without the heavy processing overhead. The GA uses the streets in the road network as its basic areal unit but any unit could be used (e.g. enumeration districts) if the connections between units could be determined.

8.2   THE SPATIAL GENETIC ALGORITHM

The accepted GA paradigm has at its core the separation of a fitness function, with all the domain specific information, from the mechanisms of selection, reproduction and generation, which are general for any application. A comprehensive introduction to Genetic Algorithms is given by Beasley (Beasley *et al.*, 1993). The GA maintains a population of encoded solutions which are given a score, by the fitness function, which is based on how well they perform on the measured criteria. New solutions are generated by combining better performing individuals though a generic set of operations (mutation and crossover) which work on the encoding. The results of these combinations form the next generation and replace the current population. Any information that is used to direct the search for better solutions is confined to the fitness function and the essential machinery for conducting the search is the same for any type of problem.

An alternative, pragmatic approach is to 'lend the GA a hand' by various problem specific optimisations. Although this has an intuitive appeal it can easily destroy the GA mechanism by removing the direct connection between the fitness function and the best solutions. If knowledge about the problem domain is to be used effectively by a GA it needs to work in conjunction with the genetic paradigm rather than an arbitrary 'bolt on' optimisation. A good account of general genetic algorithm theory and applications along with specific ideas of hybridising GAs with other techniques is given by Davis (Davis, 1991).

The task for the spatial GA is to generate typed area classifications that are contiguous and accurate. The classifications are defined by a list of roads and the accuracy is calculated by comparing the type of the houses on those roads with the type of the area classification. There are two major problems associated with this task, firstly maintaining a valid set of classifications that are contiguous and secondly, achieving a balance between number and accuracy of the areas so there are a reasonable number of reasonable size.

8.2.1 GA representation

In the spatial GA information from the problem domain is used to guide the generation of new solutions as well as providing the data to evaluate them. This is achieved by building the GA out of the relevant data structures for the problem. In essence the GA manipulates collections of roads and houses to create contiguous areas which maximise the consistency of the area classification scheme. It uses the information held by these data structures to maintain valid solutions.

Normally, the fitness function alone would be used to lead the algorithm to favour classification schemes made up of contiguous areas. However, maintaining contiguity is difficult as the combination of two valid encodings is likely to result in a discontinuous result. Without some checking operation, good solutions will produce invalid offspring rather than better solutions. In this case the encodings *are* the area classifications, so the crossover and mutation operators used to create new solutions are specifically designed to conform to the contiguity constraint and manipulate the data in meaningful ways.

8.2.2 Data structures

The implementation of the GA is divided into two separate but closely related parts. There are the genes, chromosomes and population which handle the dynamic task of creating new area classifications. Then there is the data structure for the information about the area to be investigated which holds the unchanging spatial relations and attribute data.

On initialisation, the GA reads a file that specifies the positions of the roads and the number and type of the houses that are connected to each road. A graph structure of 'arcs', representing the roads and their associated houses, is built up to create a connected network that mimics the connectivity of the original street map. The basic units of the GA are defined in terms of this map.

The gene is an area, defined by a base arc and a list of arcs that are within a certain number of connections, or steps, from the base. The link between the area and its type is provided by an intermediate level of organisation called the 'group' which contains a variable number of areas. The chromosome has one group for each of the possible types and the classification of an area is defined by its group type. There are constraints on how the genetic material is built up to ensure that the areas conform to minimum requirements regarding their size, shape and distribution.

8.2.3 The fitness function

A balance has to be found between high classification accuracy, which is likely to give overfitted, complex, solutions and solutions which sacrifice on-line training accuracy for simplicity and noise tolerance. Since each type of house needs to be represented by at least one matching area classification the simplest solution is to generate one area per class of house – better solutions will classify more houses into the correct categories. Generating more than one area per class type needs to be justified by an increased level in classification accuracy – the precise break point for

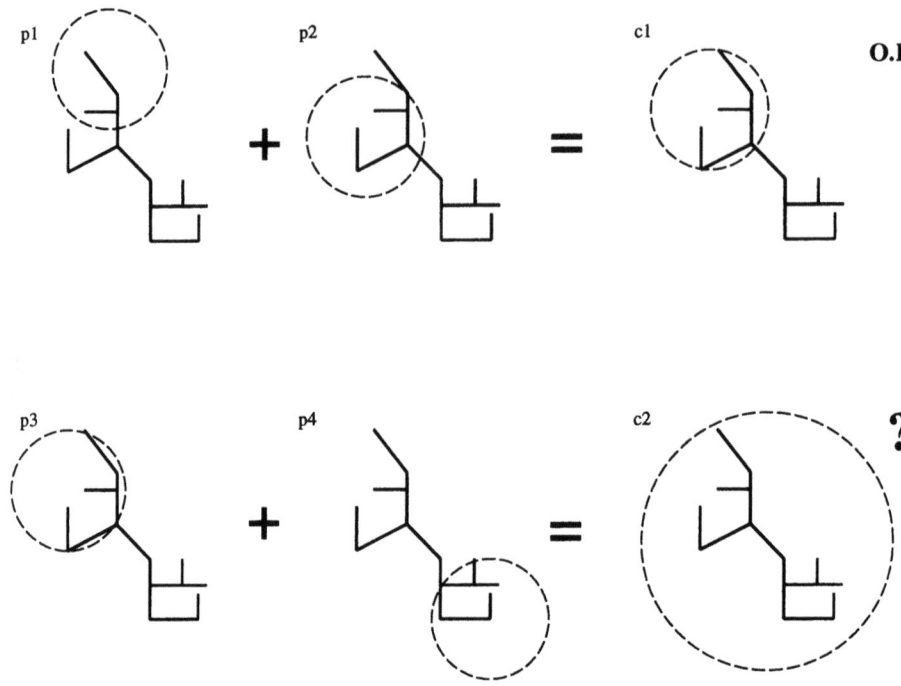

Figure 8.1 The problem of combining distinctly different areas.

this will be problem dependent. During the classification procedure each data-point falls into one of four categories:

(a) The data-point is within the area boundary and is the same class as the area – correct classification, true positives.
(b) The data-point is within the area boundary and is a different class to the area – incorrect classification, false positives.
(c) The data-point is outside the area boundary and is the same class as the area – incorrect classification, false negatives.
(d) The data-point is outside the area and a different class to the area – correct classification, true negatives.

What we want is a classification that maximizes (a) and (d) at the expense of (c) and (b). A simple way to combine these counts into a useful fitness function is to use the odds ratio (Rasmussen, 1992). which is as follows:

$$\frac{(a \times d)}{(b \times c)}$$

8.2.4   Generating solutions

Initially, a random population is generated and each area classification is evaluated so that the chromosomes can be given a fitness score. New solutions are generated by selecting individuals with a probability related to their fitness for the reproduction events of mutation or crossover. The results of these events are stored in a separate population which replaces the original when it has enough individuals. This ensures a continually changing and improving pool of potential solutions.

Crossover

There are two parts to the crossover operator, firstly the groups of the chromosomes are copied from the parents using a uniform crossover operator that gives an equal chance of inserting the groups from either parent into either child. The more complicated part comes when two groups are mixed in order to provide new, different but related offspring. A random swapping of the constituent areas will almost certainly violate the non-overlapping constraint and a random swapping of arcs between areas will be equally likely to violate the contiguous area constraint. There is a further difficulty in that it may be very reasonable to mix the information held in two very similar areas but it is difficult to find a suitable compromise between areas that are geographically distinct, as illustrated by Figure 8.1.

The solution is to combine the information provided by the whole group rather than its constituent areas. Two lists of arcs are created, one representing the union of the two reproducing groups and the other the intersection. From these lists two new groups are formed by building new areas that are created by selecting a base arc at random and adding as many of the arcs that are connected to it as possible. Areas are generated for a group until there are only one or less arcs that have not been included in an area left in the list. A final refinement is to ensure that the base arc is the most central one in each area and the minimum number of steps to include all the members is calculated. The illustration in Figure 8.2, shows how parent groups p1 and p2 crossover to provide two children. The first child is formed from all the parental arcs – the union of p1 and p2. The second child is formed from only those arcs that are common to both – the intersection of p1 and p2.

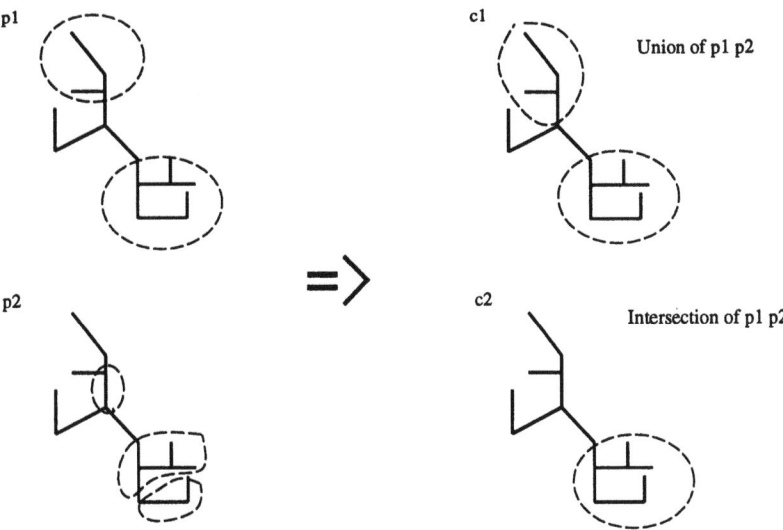

Figure 8.2 Union and intersection crossover on two area classifications.

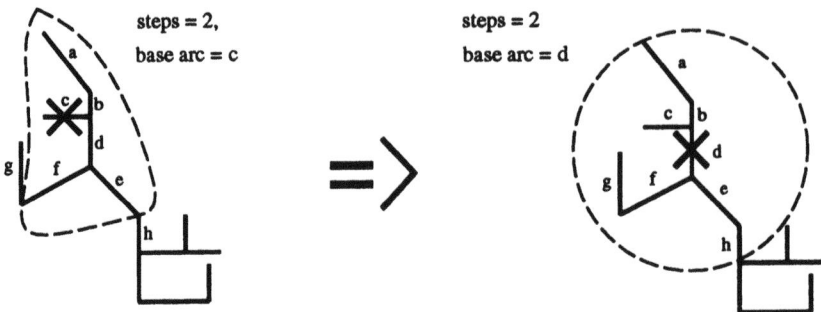

Figure 8.3 Mutation by moving the centre of an area.

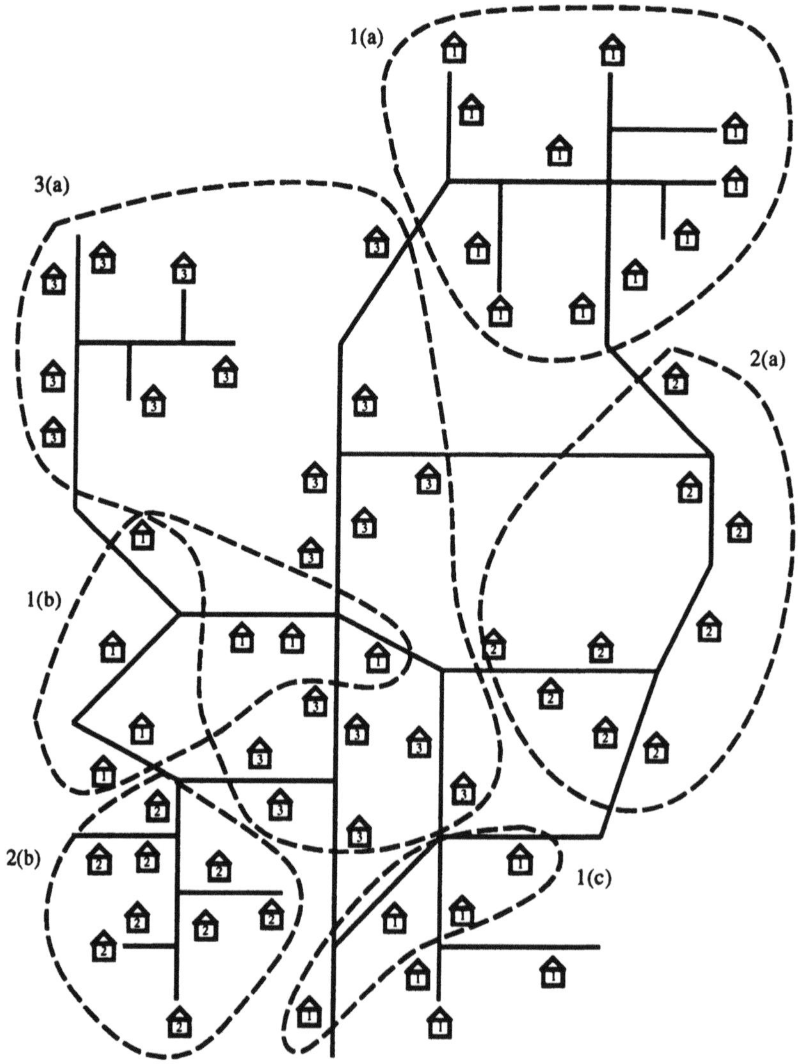

Figure 8.4 A classification showing a large, general area classification – 3(a).

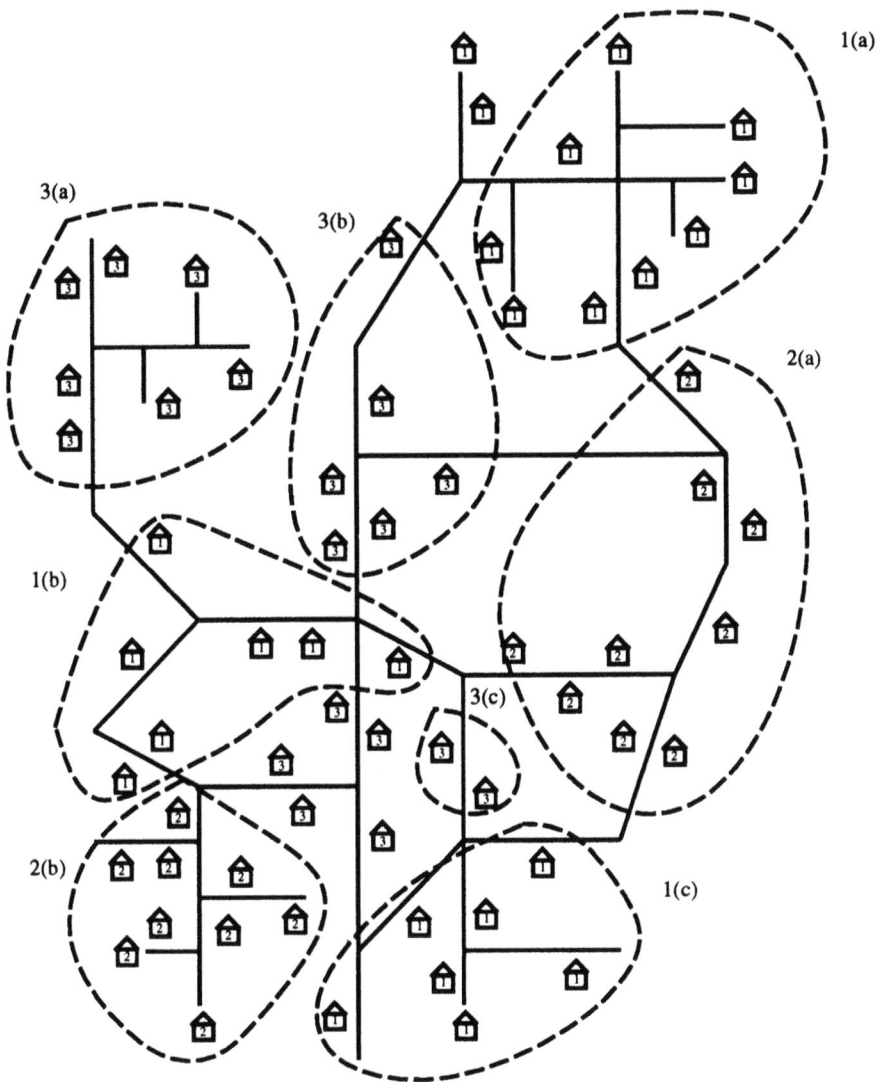

Figure 8.5 A classification showing smaller more accurate area classifications – 3(a), 3(b), 3(c).

Mutation

Due to the constraints on the problem mutation is not as disruptive as in other GA implementations. It is used to make sure that no arc can be ignored as a candidate for the base arc of an area by the GA, thus ensuring that all parts of the search space can be reached regardless of initial starting conditions. An area is mutated in one of two ways; either a neighbour of the current base arc is made the centre of the area, or an arc of the appropriate type is chosen at random, and the member arcs re-calculated.

In Figure 8.3 an area defined by the base arc 'c' and a step size of two is mutated by shifting the centre to a neighbouring arc, 'd', and although the number of steps remains the same more arcs are included into the area because of the more central location of 'd'.

8.3 RESULTS

The GA was tested on a set of data with 50 roads and 63 houses which were divided into eight distinct areas from one of three pre-determined types. From an initially random set of area classifications the GA found the optimum solution after generating and evaluating approximately a thousand area classification schemes. Also present in the final population of solutions were other classifications which did not score so highly but gave reasonable alternatives to the theoretical optimum. A previous GA implementation (Hobbs, 1995) which took a more traditional approach, typically took about six thousand evaluations to find the optimum of a spatial problem with a similar level of complexity.

Figures 8.4 and 8.5 show the mapped results of two high scoring suggestions from a run of the GA. Although the GA operates on the arcs of the road network, it is the houses which have been highlighted as these are the data points which control the classification. The features to note are that in Figure 8.4 all the class three houses have been put into the same area, 3(a), whereas in Figure 8.5 a different formulation has suggested three areas. This shows a near equivalence in fitness score between two different answers that is caused by trade-offs in the fitness function. In both figures some houses have not been classified but a composite of the two would give un-ambiguous area classifications for all the houses. The intersection of 3(a) and 1(b) in Figure 8.4 shows that the GA is simultaneously operating at multiple scales with the smaller area 1(b) invading the larger, less accurate area 3(b). Changes in the weighting of the fitness function can produce more compact, accurate, area classifications or larger, more general, classifications.

8.4 CONCLUSION

The significance of this work for GIS applications is that it provides a spatial analysis tool that is robust and friendly to users who are less experienced at using sophisticated spatial statistical methods. The GA is good for exploratory analysis as the results of the clustering are less dependent on the condition of the data than traditional statistical techniques. In addition to the best result it provides a selection of valid answers, giving extra information about the problem, which is particularly useful in the initial stages of analysis where the desired goals may be ill-defined.

The adaptation of the GA paradigm reduces the amount of computation required for the class of spatial clustering problems by tailoring a weak, generally applicable optimisation method with problem specific heuristics. Induction has been used in other spatial systems (Whigham *et al.*, 1992; Barbanente *et al.*, 1992) but the number of potential solutions that have to be generated and evaluated by these traditional implementations limits the complexity of the problems they can handle. While the example presented here is also modest in scope, it is expected that the ability of the GA to search through huge numbers of potential solutions will allow the technique to be scaled up.

The area classifications that the GA generates can be easily altered by changing the fitness function. This could be done by fine tuning the existing function by weighting the different elements, or by replacing it with a more sophisticated function. Further work is progressing to test the spatial GA on more complex spatial problems and compare the results with traditional GA and statistical methods.

REFERENCES

BARBANENTE, A., BORRI, D., ESPOSITO, F., LEO, P., MACIOCCO, B. and SELICATO, F. (1992) Automatically acquiring knowledge from digital maps by artificial intelligence planning techniques. *Theories and Methods of Spatial-Temporal Reasoning in Geographic Space*, pp. 379–401. Berlin: Springer Verlag.

BEASLEY, D., BULL, D. R. and MARTIN, R. (1993) An overview of genetic algorithms: the fundamentals. *University Computing*, **15**, 58–69.

COOLEY, R., PACK, A., CLEWER, A. and HOBBS, M. (1993) Location and residential property value. Presented at the Eighth European Colloquium on Theoretical and Quantitative Geography, Budapest.

DAVIS, L. (1991) *Handbook of Genetic Algorithms*. Morgan Kaufmann.

FOTHERINGHAM, A. and ROGERSON, P. (1993) GIS and spatial analytical problems. *IJGIS* **7**, 3–19.

HOBBS, M. (1995) Analysis of a retail branch network: a problem of catchment areas. Proceedings, 151–158. In FISHER, P., ed., *Innovations in GIS 2*. London: Taylor & Francis.

JANIKOW, C. (1993) A knowledge-intensive genetic algorithm for supervised learning. *Machine Learning* **13**, 189–228.

KIERNAN, V. (1994) Growing money from algorithms. *New Scientist* **144**, 25–28.

OPENSHAW, S. (1984) *The Modifiable Areal Unit Problem: CATMG 38*. Norwich: Geo Books.

OPENSHAW, S. (1991) Developing appropriate spatial analysis methods for GIS. In *Geographical Information Systems: Principles and Applications*, pp. 389–401. Longman Scientific and Technical.

RASMUSSEN, S. (1992) *An Introduction to Statistics with Data Analysis*. Brooks/Cole.

WHIGHAM, P., MCKAY, R. and DAVIS, J. (1992) Machine induction of geospatial knowledge. In *Theories and Methods of Spatial-Temporal Reasoning in Geographic Space*, pp. 402–417. Berlin: Springer Verlag.

CHAPTER NINE

Applying efficient techniques for finding nearest neighbours in GIS applications

MOH'D AL-DAOUD and STUART ROBERTS

9.1 INTRODUCTION

The CPU time required by clustering algorithms (e.g. k-means) is unacceptable when large data sets are involved. This is due largely to the time required to compute nearest neighbours. In this chapter, we discuss some methods to find nearest neighbours efficiently and then to implement some of these methods to clustering large data sets.

Within Geographic Information Systems (GISs), finding nearest neighbours is a common requirement. For example, spatial queries such as 'find the nearest post office to a school' obviously require a nearest neighbour algorithm. Other spatial analysis (e.g. clustering) employs nearest neighbour search extensively. For large data sets, finding nearest neighbours is a time consuming task. This problem becomes serious in clustering techniques, which is used for making decisions in the GIS realm. Clustering is used for map generalisation (Armstrong, 1991), regionalisation (Openshaw, 1983), postal services (O'Kelly, 1994), grouping economic sectors (Mulvey and Crowder, 1979) and finding optimal locations for facilities (Goodchild and Noronha, 1983). Some of the clustering techniques deal with (x, y) space, while others deal with aspatial attributes. In this chapter, we concentrate on clustering spatial (x, y) data.

One of the most popular clustering algorithms is the k-means algorithm (MacQueen, 1967). It aims to minimise the sum of the square of distances from all data points in the cluster to their nearest cluster centres. The k-means algorithm starts with initialising the clusters represented by K cluster centres. The data points are then allocated (assigned) to one of the existing clusters according to their Euclidean distance from the clusters, choosing the closest. The mean (centroid) of each cluster is then computed so as to update the cluster centre. This update occurs as a result of the movement of the cluster centres towards data points. The processes of re-assigning of the data points and the update of the cluster centres is repeated until the change in the values of all the cluster means (centroids) are within some predefined limits. However, it has been reported that the time required by the k-means algorithm to converge can become unacceptable, particularly for large clustering problems (Ismail and Kamel, 1989; Venkateswarlu and Raju, 1992). This is due largely to the amount of time required to compute nearest neighbours.

In this chapter, we discuss some solutions to this problem. We investigate the efficiency of variant methods when applied to clustering large two-dimensional data sets in (x, y) space. The methods are first tested on different data sets to find nearest neighbours. Then, they are used within the k-means clustering algorithm. The results are discussed and analysed. This chapter is organised as follows. In section 9.2 the k-d trees, quadtree and cell method are briefly discussed. The results of applying quadtree and cell method to nearest neighbours and clustering problems are reported in section 9.3. Conclusions are drawn in section 9.4.

9.2   EXISTING METHODS

Several attempts have been made to develop methods and techniques to reduce the CPU time required to find nearest neighbours. In this section, some of the most widely used methods are discussed. These are k-d trees (Friedman et al., 1977), quadtrees (Finkel and Bentley, 1974) and the cell method (Bentley et al., 1980). These methods are essentially aiming to reduce the time spent on the search for nearest neighbours. Also in this section, a modification of the cell method is presented. This modification leads to further savings in CPU time when applied to clustering problems.

9.2.1   K-d trees

Friedman et al. (1977) suggested the use of k-d trees (Bentley, 1975) to find nearest neighbours. The k-d tree is a multidimensional generalisation of binary trees. The root of the k-d tree represents the k-space containing the set of all cluster centres. Each node represents a subset of cluster centres. Each non-terminal node has two son nodes (left and right sons). These son nodes are obtained by partitioning the node's subset into two parts by a hyperplane orthogonal to one of the k-co-ordinate axes (called discriminating axis). The position of the hyperplane is chosen so that each son contains approximately half of the parent's cluster centres. The discriminating axis is determined by chosing the most dispersed co-ordinate, that is, the variance is maximised. Sons that contain less than a predefined threshold number of cluster centres form the terminal nodes of the tree. Each of these terminal nodes defines a cell in the multidimensional space containing a subset of cluster centres, with cell size less than the predefined threshold.

The search for nearest cluster centre for a data point starts by descending through the tree, deciding at each non-terminal node to investigate either the left or right son according to whether the data point lies on the left or right side of the partitioning plane, until a terminal node is found. If a ball centred at the data point with radius equal to the distance between the data point and current nearest cluster centre does not overlap the cell's boundary, then all cluster centres in the cell represented by the found terminal node are exhaustively tested for nearest and the search is stopped. However, if the ball overlaps the cell's boundary, backtracking is required and the search for nearest must be performed in the opposite side of the partition in which case the computation complexity is increased (Sproull, 1991). Applying the k-d trees to clustering has another problem. Because the problem is one of finding the nearest cluster centre to a given data point (rather than vice versa), there is an overhead related to re-building (updating) the tree after each clustering iteration.

9.2.2 Quadtrees

The quadtree (Finkel and Bentley, 1974) is a hierarchical data structure which recursively subdivides the plane into blocks based on some decomposition rule (usually a threshold on the number of points). The decomposition process starts from a square that contains all cluster centres to be represented, and proceeds with a recursive subdivision into four equal-sized quadrants. Corresponding to this subdivision is a tree structure, in which each node of the tree has four descendents.

Finding the nearest cluster centre to a given data point is done by finding the smallest circle that encloses the data point and any cluster centre; the minimum distances between the data point and all cluster centres in the circle are then computed to find the nearest.

9.2.3 The cell method

The cell method (Bentley *et al.*, 1980) is a technique in which the distance from a point to the nearest cluster centroid is calculated in expected constant time, independent of data size. The idea of the cell method is to divide the data space into small squares (called cells) and to associate with each cell a pointer to a list of all cluster centres in that cell.

To find the nearest cluster centre to a data point, a search is performed, starting at the cell holding the data point and searching, in a spiral-like pattern, the cells surrounding it, until a cluster centre in a non-empty cell is found. Once one cluster centre is found, it is guaranteed that there is no need to search any cell which does not intersect the circle of radius equal to the distance to the first cluster centre found and centred at the data point. Thereafter, only cells that intersect the circle centered at the data point with radius equal to the distance to the found cluster centre need to be searched. In this way the nearest cluster centre can be located in constant expected time. We expect relatively small overheads for the allocation of cluster centres to cells since this allocation is straightforward.

9.2.4 Enhancement of the cell method

The procedure of finding a non-empty cell and determining the circle is done repeatedly for each data point in the cell. For example, if the cell holds M data points, then M number of circles must be determined, which is time consuming. Instead, only one circle might be determined for all data points in the cell centered at the cell center and only one search is required to find cluster centres in that circle. Experiments in this study showed that up to 16 per cent of the time could be saved when applying the above technique to non-uniform data.

9.3   EXPERIMENTAL RESULTS

Table 9.1 shows the expected time for each of the above methods to compute nearest neighbours. For example, the expected time for both k-d trees and quadtrees is (log N), while the expected time is constant for the cell method. However, as shown in Table 9.1, each of the methods has a drawback. In k-d trees and quadtrees backtracking might occur in the process of finding nearest neighbours, while, in the cell method, if the data is distributed non-uniformly, its efficiency decreases. Since k-d trees and quadtrees have similar behaviour, quadtrees are implemented together with the cell method to both nearest neighbours and clustering problems. In this section, the quadtrees and cell methods are first implemented to compute nearest neighbours. Then, they are implemented to solve clustering problems.

9.3.1   Applications to nearest neighbour search

Our approach is initially to investigate the efficiency of both quadtrees and cell methods for speeding up the process of finding nearest neighbours. These two methods are implemented to two different data sets. Two experiments have been performed. The purpose was to compare the run time taken by each method to compute nearest neighbours, ignoring the 'set up' time for building the index structure. The run time relating to an exhaustive search is also computed. For both experiments, a set of 1000 query points was generated randomly.

In the first experiment, five different sets of data points were generated randomly. Then, for each query point, the nearest data point was identified. The results are shown in Figure 9.1. The Figure shows that the cell method performed the best, especially when the number of data points increases.

In the second experiment five new sets of data points were generated non-uniformly. The nearest point was identified to each of the query points. The results are shown in Figure 9.2. The Figure shows that the cell method performed the best. However, it was noticed that the performance of the cell method decreased when using data with a non-uniform distribution (see Figure 9.3 for comparison). This degradation is because the cells are either very dense or sparse, which leads to more time being spent on the search processes.

On the other hand, the results show that quadtrees have almost the same performance when the data is either randomly or non-uniformly distributed, which is

Table 9.1 Comparisons of the characteristics of some existing methods for finding nearest neighbours

	Cell	k-d trees	Quadtrees	Exhaustive
Time	Expected constant	$O(\log N)$	$O(\log N)$	$O(N)$
Drawback	Non-uniform data	Backtrack	Backtrack	$O(N)$

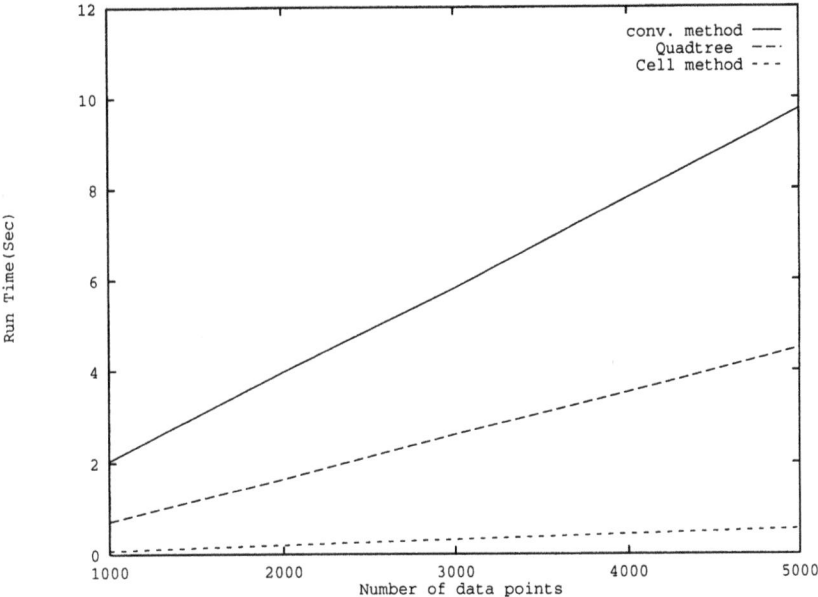

Figure 9.1 Comparisons of different methods to find nearest points (randomly distributed).

an advantage of quadtrees over the cell method (see Figure 9.3). It is not clear whether the cell and quadtree methods will give good results for clustering since they impose an overhead which will be repeated for each iteration of the k-means algorithm. This is discussed in the next section.

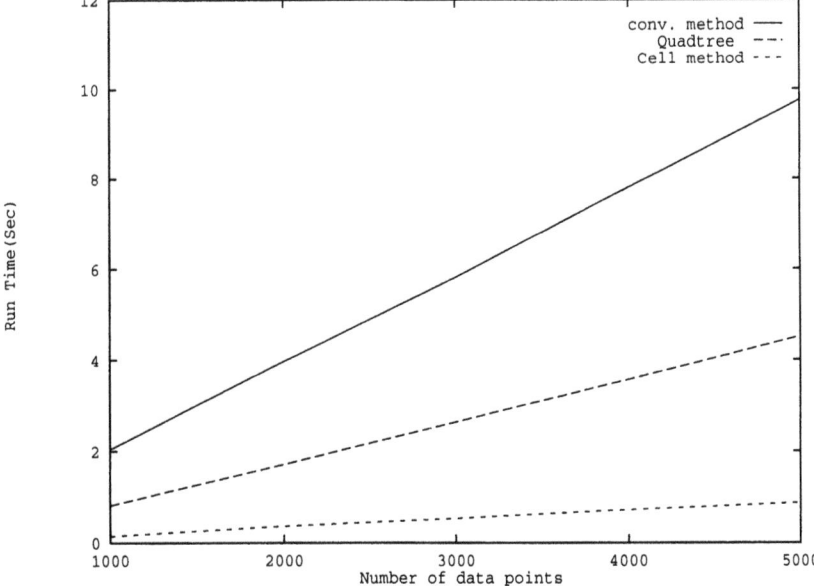

Figure 9.2 Comparisons of different methods to find nearest points (non-uniformly distributed).

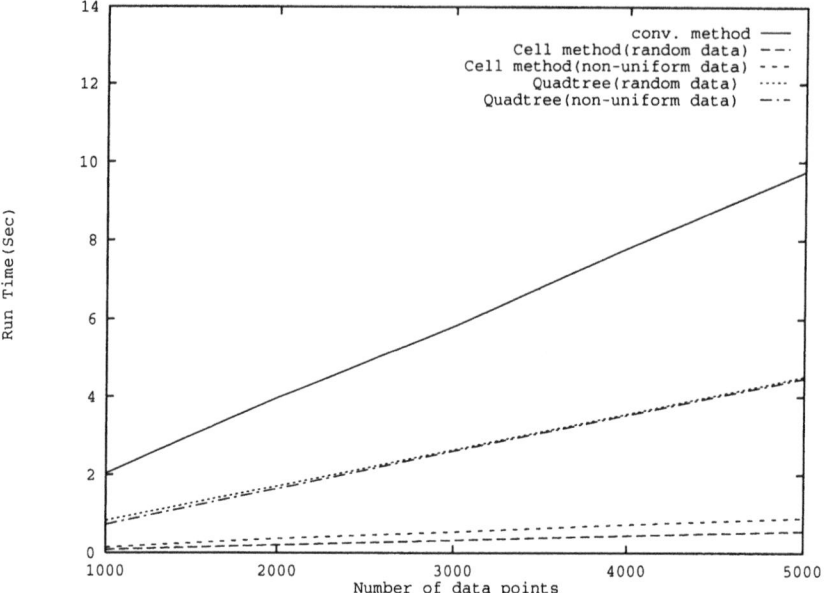

Figure 9.3 The behaviour of different methods for random and non-uniform data sets.

9.3.2 Applications to k-means clustering algorithm

To determine the performance of the methods when applied to the k-means clustering algorithm, several experiments have been conducted with different data sets. The purpose was to compare the run time of the k-means when the quadtrees and cell methods are applied on the one hand and when the conventional method (i.e. exhaustive search) is applied on the other. In these experiments, the same initial configuration has been used for all methods, which guarantees that all methods converge to the same solution. One of these experiments was to cluster a set of data points (which contained about 50 000 data points). These data points correspond to positions of water hydrants in an area including the city of Leeds. The data is highly non-uniformly distributed.

The performance of the methods is shown in Figure 9.4. The Figure shows that the cell method outperformed the conventional and quadtree methods. Unfortunately, Figure 9.4 shows that the use of quadtrees leads to poor performance. This is because of the overhead of re-building (updating) quadtrees, which is very time consuming, while this overhead is very small in the cell method because the allocation of cluster centres to cells is straightforward.

Another experiment was to cluster a set of data points (which contained about 7500 data points). These data points correspond to positions of selected households in one of the areas of Leeds. This data set was chosen because of the highly non-uniform distribution of points. The performance of the methods is shown in Figure 9.5. The Figure shows that again the cell method showed greatest efficiency. On the other hand, the results showed that the time spent on clustering has been significantly reduced using the cell method for large (>40) number of clusters. In addition, the improvement in the performance of the cell method continues to increase when the number of clusters increases.

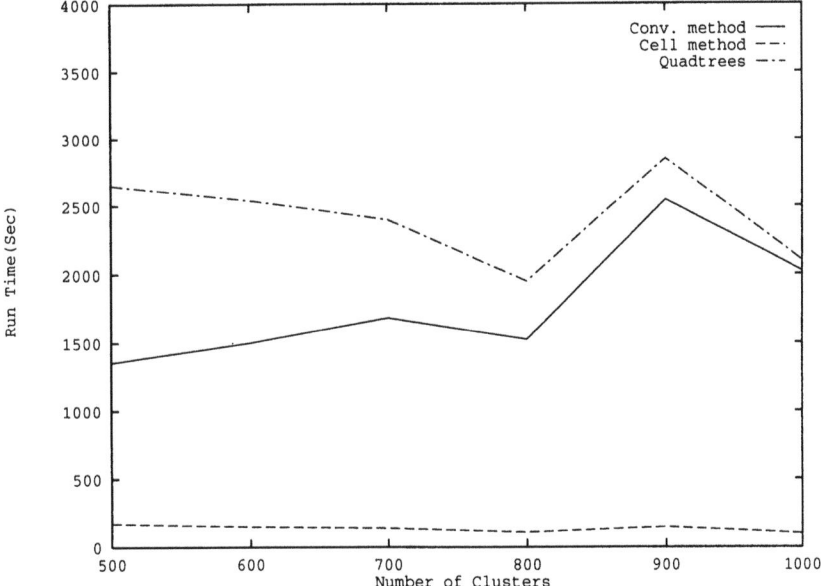

Figure 9.4 The performance of the conventional, cell and quadtree methods to cluster the hydrant data.

An obvious objection to these results is that we have 'tuned' the cell size to give optional performance and that for real clustering problems the optimal cell size would be unknown. However, experiments show that the method is not unduly sensitive to the size of the cell and that the optimal cell size corresponds to about

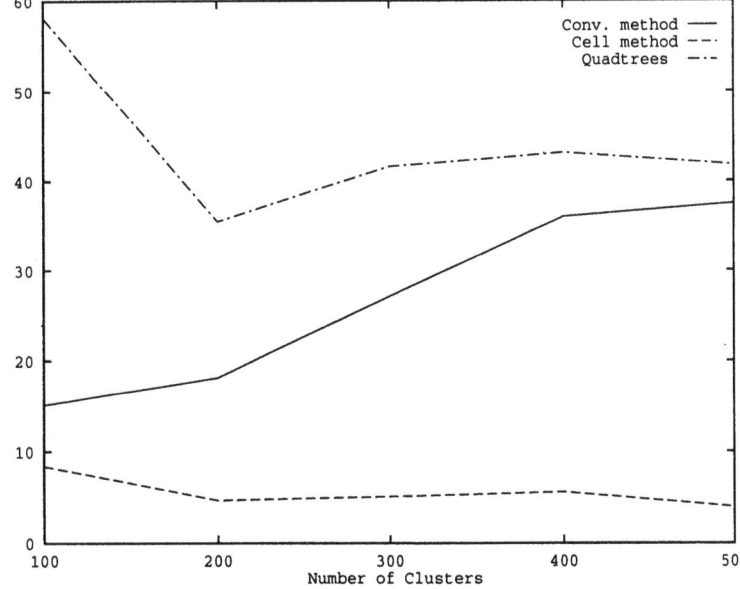

Figure 9.5 The performance of the conventional, cell and quadtree methods to cluster the households data.

three cluster centres on average in each cell for all sets of data for which the methods were applied.

9.4 SUMMARY AND CONCLUSIONS

Clustering is used in GIS to help geographers in making decisions. However, clustering algorithms require a large amount of computation time to find nearest neighbours which hinders the development of an effective GIS. In order to minimise this computation time, some methods have been discussed in this chapter.

The experiments that have been performed on clustering processes have shown that the cell method performs efficiently when using data with large (>40) number of clusters. Although the performance of the cell method decreases when the data is non-uniform, it nevertheless represents an efficient tool for computing nearest neighbours and clustering for each of the data sets considered.

In addition, our experiments showed that the cell method in not sensitive to the size of the cell. However, it has been found that the optimal number of cluster centres in each cell is about three on average. Further, we have succeeded in enhancing the cell method so that up to 16 per cent of the CPU time could be saved.

For data distributed non-uniformly it was expected that quadtree division would provide better efficiency than uniform square cells. Some experiments have been performed to test this hypothesis but indicate that the overheads involved in maintaining the quadtrees structures outweigh the advantages. The cell method should provide efficient and practical tools for GIS.

REFERENCES

ARMSTRONG, M. (1991) Knowledge classification and organization. In BUTTENFIELD, B. and MCMASTER, R., eds, *Map Generalization: Making Rules for Knowledge Representation.* Longman, pp. 86–102.

BENTLEY, J. (1975) Multidimensional binary search trees used for associative searching. *Comm. ACM* **18**, 509–517.

BENTLEY, J., WEIDE, B. and YAO, A. (1980) Optimal expected-time algorithm for closest point problems. *ACM Trans. Math. Soft.* **6**, 563–580.

FINKEL, R. and BENTLEY, J. (1974) Quadtrees: a data structure for retrieval on composite keys. *Acta Inform.* **4**, 1–9.

FRIEDMAN, J., BENTLEY, J. and FINKEL, R. (1977) An algorithm for finding best matches in logarithmic expected time. *ACM Trans. Math. Soft.* **3**, 209–226.

GOODCHILD, M. and NORONHA, V. (1983) *Location-Allocation for Small Computers.* Monograph no. 8, Department of Geography, University of Iowa, Iowa City, IA.

ISMAIL, M. and KAMEL, M. (1989) Multidimensional data clustering utilization hybrid search strategies. *Pattern Recognition* **22**, 75–89.

MACQUEEN, J. (1967) Some methods for classification and analysis of multivariate observations. Proceedings of the 5th Berkeley Symposium on Mathematics, Statistics and Problems, 281–297.

MULVEY, J. and CROWDER, H. (1979) Cluster analysis: an application of Lagrangian relaxation. *Management Sci.* **25**, 329–340.

O'KELLY, M. (1994) Spatial analysis and GIS. In FORTHERINGHAM, S. and ROGERSON, P. eds, *Spatial Analysis and GIS.* London: Taylor & Francis, pp. 65–79.

OPENSHAW, S. (1983) Multivariate analysis of census data: the classification of areas, In RHIND, D. ed. *A Census User's Handbook*, Methuen, pp. 243–263.

SPROULL, R. (1991) Refinements to nearest-neighbor searching in K-dimensional trees. *Algorithmica* **6**, 579–589.

VENKATESWARLU, N. and RAJU, P. (1992) Fast isodata clustering algorithms. *Pattern Recognition* **25**, 335–342.

CHAPTER TEN

Computational techniques for determining three-dimensional topology

HUGH BUCHANAN

10.1 BACKGROUND

Topology is held and transferred between systems for three main reasons:

- to simplify spatial queries, by storing connectivity explicitly, and thereby reducing a query from a geometric calculation to looking up a table;
- to ensure consistent analysis of spatial relationships, by avoiding the difficulties of computational precision, and
- reducing the effort required by a data recipient in making a received data set usable in their system.

It is always possible to generate topology from geometry, although, as noted above, different people may derive a different topology, depending on the computational method and precision used. It may also be useful to validate the stated topology in a set of data. This validation consists of ensuring that the stated topological relationships are consistent with the given geometric positions. A validation process such as this has great significance in testing conformance of data sets with published transfer standards (which often incorporate topological rules).

Topology is often taken for granted, without considering how these relationships have been determined by calculations on geometric data. This paper looks at one possible set of geometric calculations that can be used to determine topology. In particular, we look at these calculations in three-dimensional space, where both the geometry and the topology are more complex than in two-dimensional space. This area has been previously studied by those working in the field of solid modelling, among others by Preparata and Shamos (1985) and Mantyla (1988). The approach of this paper differs in certain computational approaches, and by making the material accessible to the wider geographical data community by avoiding extensive mathematical terminology.

The terminology that we will use is that a node is a zero-dimensional topological primitive, an edge is a one-dimensional topological primitive, a face is a two-

Table 10.1 The ten cases of interaction between topological primitives

	Node	Edge	Face	Cell
Node	x			
Edge	x	x		
Face	x	x	x	
Cell	x	x	x	x

dimensional topological primitive, and a cell is a three-dimensional topological primitive.

10.2   THREE DIMENSIONS

Working in three dimensions presents challenges which are significantly different from those involved when working in two dimensions. In terms of topology, more topological components can legitimately lie within each other. For example, it is legitimate to have a node or edge on a face, or a node, edge or face within a cell (Rikkers *et al.*, 1994). The complete set of relationships that need to be considered between topological primitives of different dimensions is shown in Table 10.1.

In addition, in both the two- and three-dimensional cases, stated topological relationships can be used to supplement given geometric position. For example, with a coordinate resolution of 1 m, consider a face with vertices at $(0, 0, 0)$, $(3, 0, 2)$ and $(0, 3, 2)$, and point at $(1, 1, 1)$. Simple coordinate geometry shows that the node lies off the face by 0.3 m. If it was stated that the node in fact lies on the face, and it was known that the stated topology was correct, then this would refine the geometric description of the face. This problem does not arise to the same extent in two-dimensional topology, since the only topological containment that is permitted is node or edge on face (DGIWG, 1994). These two aspects make an understanding of three-dimensional geometric calculations very important and make their robust implementation necessary.

We are considering objects that exist in three-dimensional space. The methods developed for these tasks rely on determining the intersection of node, edge, face and cell components. In each case, we need to consider the intersection of each primitive with each primitive of equal or lower dimension, as shown in Table 10.1. For simplicity, we will consider only edges that are straight line segments, and faces that are triangular. The restriction of a face to be a triangle is necessary in order to make its definition unambiguous. Any three points in three-dimensional space define a plane, and the face is taken to be the triangle formed in that plane by these points. If the face is allowed to have more than three vertices, then it cannot be guaranteed that the vertices all lie in the same plane. If the vertices do not lie in the same plane, then the definition of the interior of the face is ambiguous. In other texts, this potential difficulty is often not directly addressed, and the methods developed allow faces with any number of vertices.

10.3 NATURE OF CELLS

It might be assumed that each cell needs to be a tetrahedron (a tetrahedron is a solid figure bounded by exactly four triangular faces, as shown in Figure 10.1), by analogy with the triangular form that is used for simple faces.

As noted above, the need to restrict faces to triangles arises to force each face to lie in a two-dimensional space (a plane) in order for them to be used to define an unambiguous boundary of a three-dimensional object. A volume always forms an unambiguous boundary in three-dimensional space. There would only be a need to restrict volume elements to tetrahedra if they were to be used as the boundary for an object in a four-dimensional space. In this case, each vertex would be described by four co-ordinates. While we are working in three-dimensional space, we can allow cells to be of a general shape, but with triangular faces. An additional rule for the faces of a cell is the need for them to have their boundary consistently defined in either a clockwise or anticlockwise direction when the volume is viewed from the outside.

Our cells therefore have any number of faces, but each of these faces is triangular. It is useful to be able to state the relationships between the number of vertices, edges, and faces in such cells, denoted by V, E and F respectively. We start from the well-known Euler's equation (for example, Preparata and Shamos (1985)) which states that $F + V = E + 2$. This remarkably simple identity is true for all three-dimensional objects with plane faces.

For cells with triangular faces, a stronger set of identities can be derived, which are:

$$3V = E + 6$$

$$3F = 2E$$

$$2V = F + 4$$

$$E > F > V \text{ since } V \geq 4 \tag{10.1}$$

The derivation of these identities follows from consideration of the process of removal of a single vertex (we call it R) from a cell. Say that R is located at the end of n edges, and is hence directly connected to n adjacent vertices, and is a vertex on n faces. The removal of R results in an n-sided polygonal face whose vertices are the

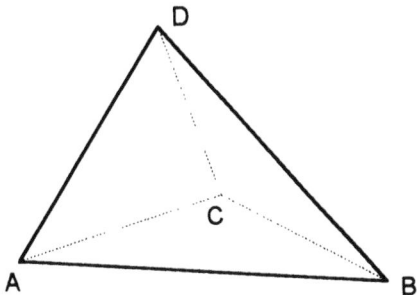

Figure 10.1 A tetrahedron.

Table 10.2 Reduction in number of vertices, edges and faces as vertices of a polyhedron with triangular faces are removed

Number of vertices	Number of edges	Number of faces
V	E	F
V-1	E-3	F-2
....	....	....
4	6	4

adjacent vertices of R. To conform with our definition of a cell, this n-sided polygon must be divided into triangles. This results in the creation of n-3 new edges, and n-2 new triangular faces. This allows us to consider the effect of the removal of R on each of V, E and F in turn. Most simply, V is reduced by one. For E, n-edges are removed, but n-3 new edges created, resulting in a net decrease of 3. Similarly for F, n faces are removed, but n-2 new faces created, resulting in a net decrease of 2. If we additionally note that for a tetrahedron $V = F = 4$ and $E = 6$, we can create the simple Table 10.2.

By considering the number of steps to go from the first line to the last line in Table 10.2, we can write $3(V - 4) = (E - 6)$, and $2(V - 4) = F - 4$ from which the identities in equation 10.1 follow.

Tetrahedra are convenient for computation and therefore, although we are allowing more general cells, it is useful to have a mechanism for converting a cell into a series of tetrahedra. A simple method for this is given by:

- select a vertex, say C.
- form a tetrahedron with C and each face of the cell in turn.
- the resulting tetrahedron is treated as positive if C 'sees' the inside of face, and negative otherwise.

The original volume is treated as the signed sum of these tetrahedra. Any point that lies within the volume will lie within more positive tetrahedra than negative. Special care has to be taken with points that lie on the boundary of tetrahedra, in the same way that special care has to be exercised when using trial lines for a point in the polygon test.

This approach is robust in that it can cope with volumes that have holes, and is computationally simple. It results in F tetrahedra, if we include the trivial tetrahedra formed by those faces adjacent to C.

10.4 BOXING TESTS

Each test for intersection of topological primitives involves a significant computational effort. It is therefore useful to use an approximate test that is computationally simple, and will discover most of the cases where no intersection between two primitives is possible. A suitable test for this is a boxing test where the minimum and maximum extent, in each x, y and z, of each object being tested are used to generate a cuboid. If these two cuboids do not intersect, then the two primitives cannot intersect. Such tests are well documented in the two-dimensional case (for example, Newman and Sproull (1979)). Their suitability depends on the fact that intersection of the cuboids can be determined by simple inequality tests between the relevant bounds, which are computationally very easy to carry out. Such tests will produce a number of 'false positives', where the two cuboids intersect, but the two primitives do not intersect. It will however not produce any 'false negatives', where the primitives intersect, but the cuboids do not.

The simple two-dimensional test can be refined by subdividing the original box into a series of smaller boxes. This reduces the number of false positives, and hence the precision of the test. The same refinement can be applied in the three-dimensional case, but only to edges. A face fills its enclosing cuboid to a much larger extent, and hence the cuboid cannot be conveniently subdivided, while still containing the original face.

To try to determine the optimum amount of refinement, we can determine a relationship between the amount of computing time required (T), and the number of subdivisions of the original cuboid (n). We will use the following notation:

V = volume of cuboid
p = probability of intersection per unit volume
t_c = computing time for boxing test
t_g = computing time for full geometrical test
n = number of subdivisions along a side of the cuboid

$$\text{time saved by detecting false positives} = t_g pV(1 - n/n^3)$$

$$\text{time spent in doing so} = t_c n^2$$

$$\text{computation time } T = t_c n^2 - t_g pV(1 - 1/n^2) \tag{10.2}$$

We wish to find a minimum for T as n varies, and so:

$$\frac{\partial T}{\partial n^2} = t_c - t_g pV(1/n^2)^2$$

$$\text{minimum for } T \text{ when } \frac{\partial T}{\partial n^2} \text{ is } 0 \Rightarrow (1/n^2)^2 = \frac{t_c}{t_g pV}$$

$$\Rightarrow n^2 = \left(\frac{t_g pV}{t_c}\right)^{1/2} \tag{10.3}$$

Thus, we have an optimum value for n, which is expressed in terms of the known quantities t_c, t_g and V and the unknown quantity p. In practice, p will vary between datasets, and between regions within a dataset, and cannot be determined in advance. The most appropriate approach is to determine p empirically during execution.

Secondly, although we have treated n as a continuous variable, it can only take integer values, and therefore the expression for T (equation 10.2) needs to be evaluated for the integer values of n immediately above and below the value determined in equation 10.3.

10.5 EDGE–FACE INTERSECTION

One interaction that demonstrates the type of method we are discussing, and that is useful in determining several other interactions is the intersection of an edge (a

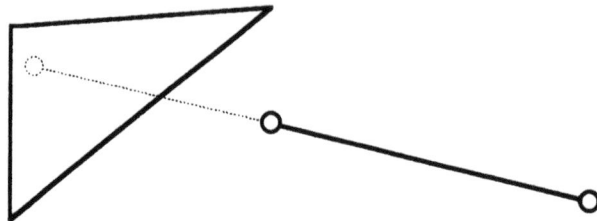

Figure 10.2   Intersection of line segment and triangular face.

straight line segment) with a triangular face. This situation is illustrated in Figure 10.2.

The technique developed here first involves extending the triangular face to a complete plane, and extending the edge to an infinite straight line. Having done this, the line and the plane will meet at a single point, except in the special case where they are parallel. It is straightforward to determine if this point lies within the line segment (Bowyer and Woodwark, 1983). To determine whether the point of intersection also lies within the triangular face requires a little more work.

We transform from three-dimensional co-ordinates to a two-dimensional co-ordinate system whose axes are defined by two sides of the triangle. This co-ordinate space contains the triangular face and the intersection point. By thus reducing the dimensionality of the problem, we are then left with a straightforward two-dimensional point in triangle calculation. Details of this method are given in the Appendix. The importance of this technique is in its applicability to many of the tests for intersection of other topological primitives.

The approach taken here can be compared with the approach taken by Mantyla (1988). Mantyla projects the face concerned onto one of the xy plane, the xz plane or the yz plane. He selects which of the planes to use by taking the largest element of the normal vector to the face and eliminating that component. For example, if the triangular face is normal to the vector $(0, 0, 1)$ then the z element is largest and the face would be projected onto the xy plane.

Mantyla's approach has the disadvantage of scaling the x, y and z co-ordinates of the original space each by different amounts during the projection process. This makes the reliable application of tolerance values more difficult. Such tolerance values are always necessary in computational tasks, but may be especially important in handling data of variable spatial accuracy. The approach of this paper has the significant disadvantage of computational intensity.

10.6   FACE–FACE INTERSECTION

There are two possible ways of approaching this problem. The first relies entirely on the calculation of edge–face intersections. It uses the fact that two faces can only intersect either if two edges of one face intersect the other face, or one edge of each face intersects the other. These two possible situations are illustrated in Figure 10.3. Both of these situations can be tested for by repeated application of the edge–face test. The test needs to be applied a maximum of five times. If no intersection is found in any of these cases, then none exists.

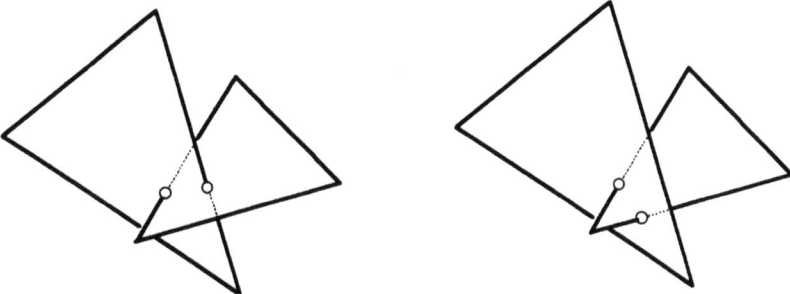

Figure 10.3 The two possible cases of intersection of two triangular faces.

The alternative approach to the intersection of faces is to calculate the infinite line of intersection between the planes of the two faces. Once this line has been determined, then it can be tested (as a two-dimensional problem) for intersection with each of the faces. If the line does intersect both faces, then a further test is required to determine whether the sections of the line that meet each face have any points in common with each other. This latter approach is more direct, but is computationally more lengthy.

10.7 NODE–CELL INTERSECTION

One simple approach to determining whether a node lies inside a cell is to decompose the cell into a series of tetrahedra, as described previously. The node lies within the cell if it lies within more positive tetrahedra than negative tetrahedra. To test whether a node lies within a tetrahedron is straightforward, if the parametric equation of the tetrahedron is used. In this form of equation the tetrahedron is described by equation 10.4.

$$\begin{pmatrix} x \\ y \\ z \end{pmatrix} = \begin{pmatrix} xA \\ yA \\ zA \end{pmatrix} + s\begin{pmatrix} xBA \\ yBA \\ zBA \end{pmatrix} + t\begin{pmatrix} xCA \\ yCA \\ zCA \end{pmatrix} + u\begin{pmatrix} xDA \\ yDA \\ zDA \end{pmatrix}$$

$$s \geq 0, \, t \geq 0, \, u \geq 0, \, s + t + u \leq 1 \quad (10.4)$$

To determine whether a node lies within a tetrahedron, the x, y, and z co-ordinates of the node can be converted to co-ordinates in s, t, and u, and then tested to see whether they conform to equation 10.4.

Alternatively, a test can be constructed that determines whether the node lies on the inside or outside of each face of the cell. This test uses the fact that the direction of the perimeter of faces has been consistently defined, and calculates a scalar triple product between the vector connecting the node to the face, and two edges of the face. The sign of this triple product indicates the side of the face on which the node is situated.

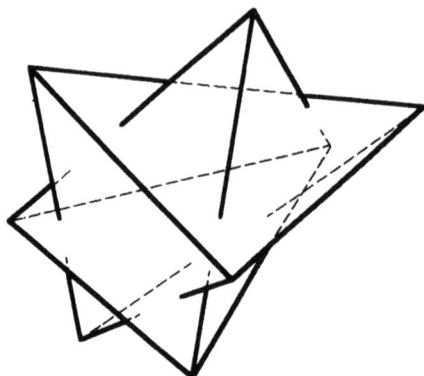

Figure 10.4   Two intersecting tetrahedra.

10.8   CELL–CELL INTERSECTION

The most complicated interaction between topological primitives is the intersection of cells. The testing of this requires extensive calculation. It is not sufficient to check simply for the containment of the vertices of one cell within the other, since the two cells can intersect, without any vertices being contained in the other cell. This can be visualised in Figure 10.4 showing two tetrahedral cells which intersect in this way.

The simplest approach involves looking for intersections between the faces of the two cells, by testing all the faces of one against all the faces of the other. This involves repeated application of whichever of the methods described above was preferred. This needs to be supplemented by a test of one vertex of each cell for containment in the other cell, to catch the case where the one cell completely contains the other.

10.9   CONCLUSIONS

There is a need for a robust, simple set of tools for deriving topology from geometric co-ordinates, or validating whether the stated topology reflects the underlying geometry. This objective is distinct from the aims of solid modelling where the emphasis is more on the dynamic creation of models by addition, deletion or movement of points, lines, planes and volumes. These two areas are closely related, and the book by Mantyla gives an excellent description of the processes of solid modelling.

Computational efficiency is one factor to be considered when selecting a suitable method. The use of refined boxing tests is one way of enhancing this. We have illustrated the approach taken for exact tests by describing a test for the interaction between an edge and a face.

Data models for describing full topological relationships are becoming better developed, as described in Rikkers *et al.*, (1994). This will increase the need for reliable methods for determining the validity of stated topology, and for constructing correct topology from three-dimensional geometry.

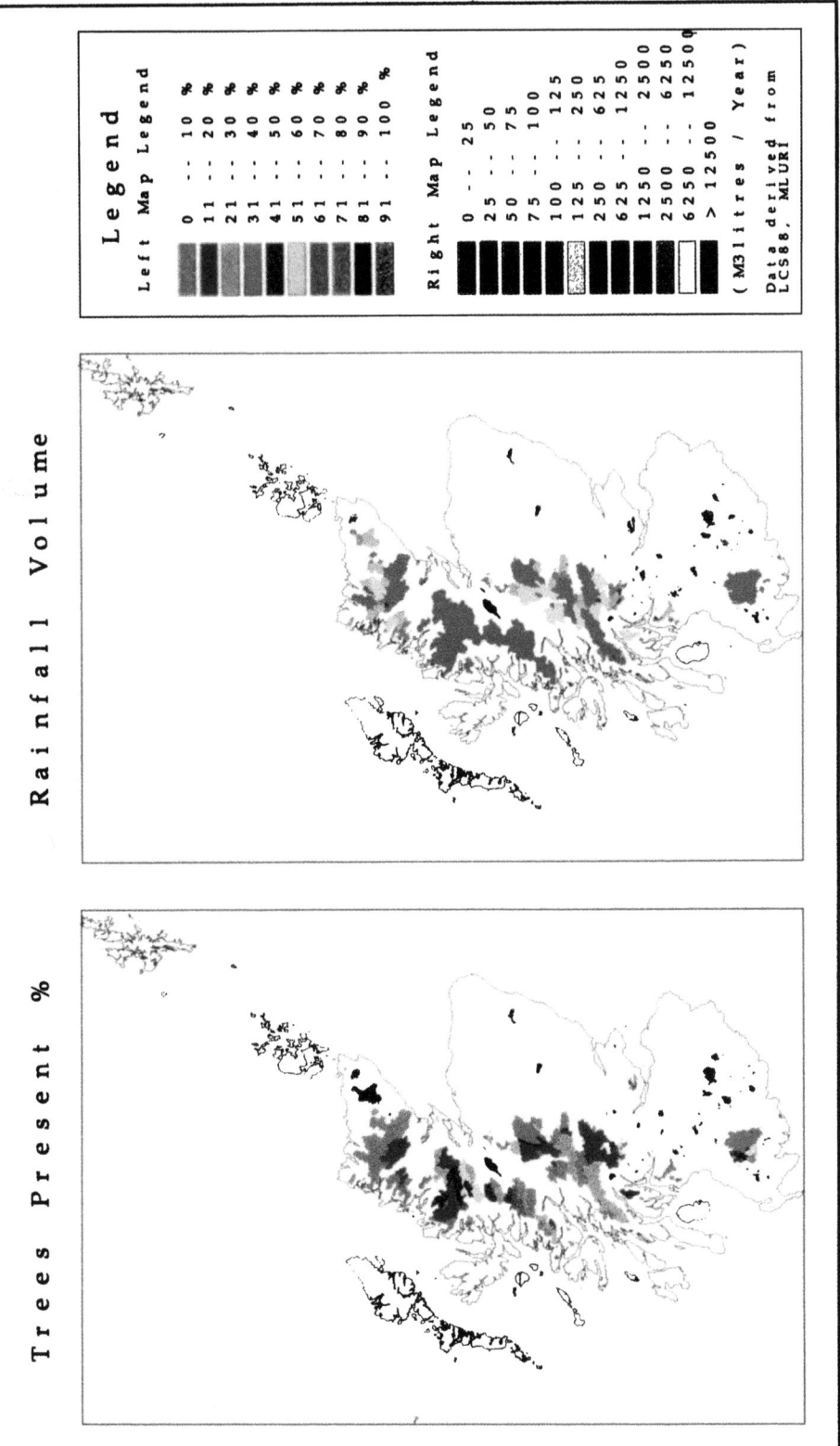

Plate 1 Environmental characteristics of catchments within a catchment database. *Left*: percentage of land which has trees present. *Right*: annual volume of input rainfall to catchments.

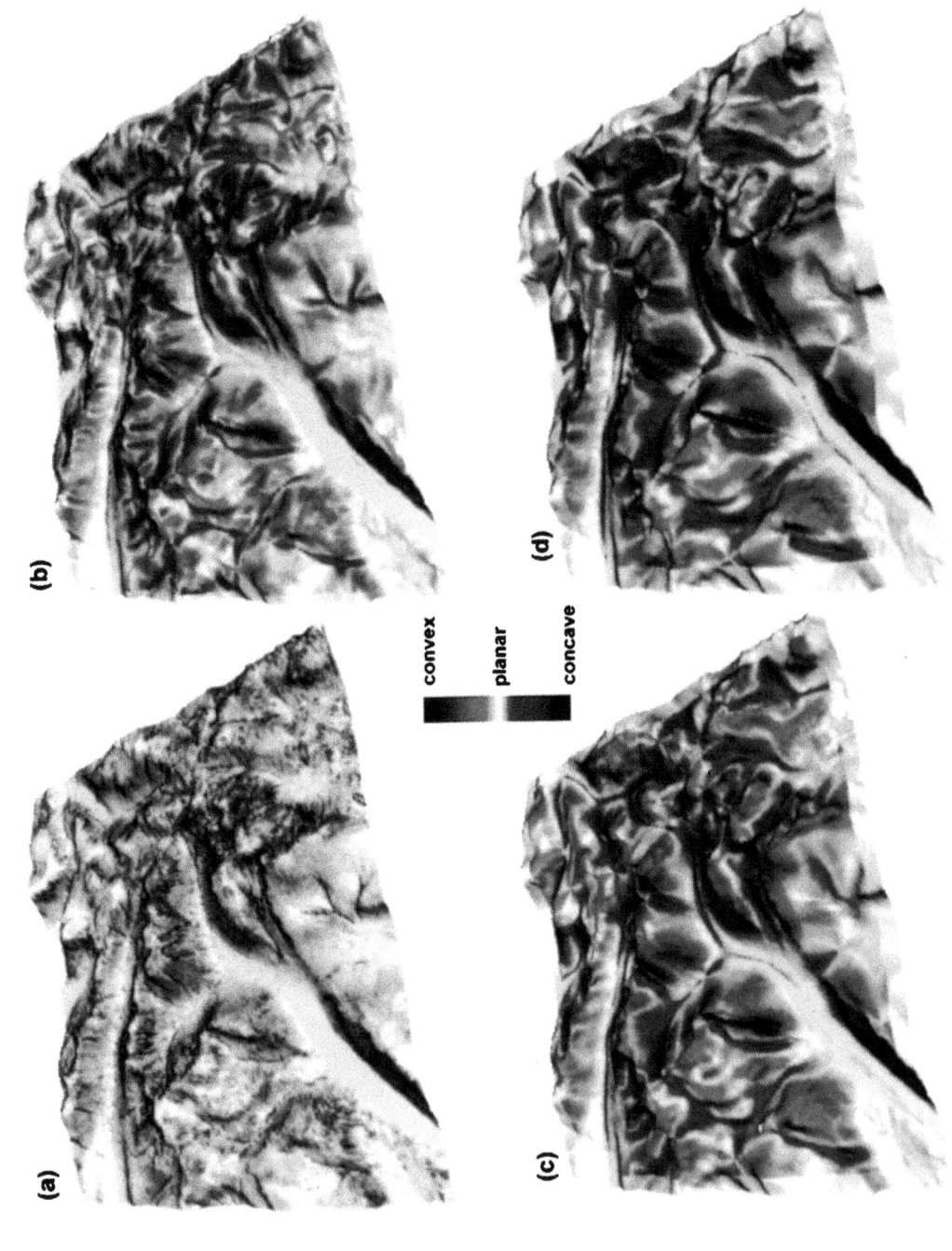

Plate 2 Visualisations of surface curvature at four kernel scales ((a) 150m; (b) 450m; (c) 850m; (d) 1250m). Darker blue indicates greater concavity, darker red indicates greater convexity.

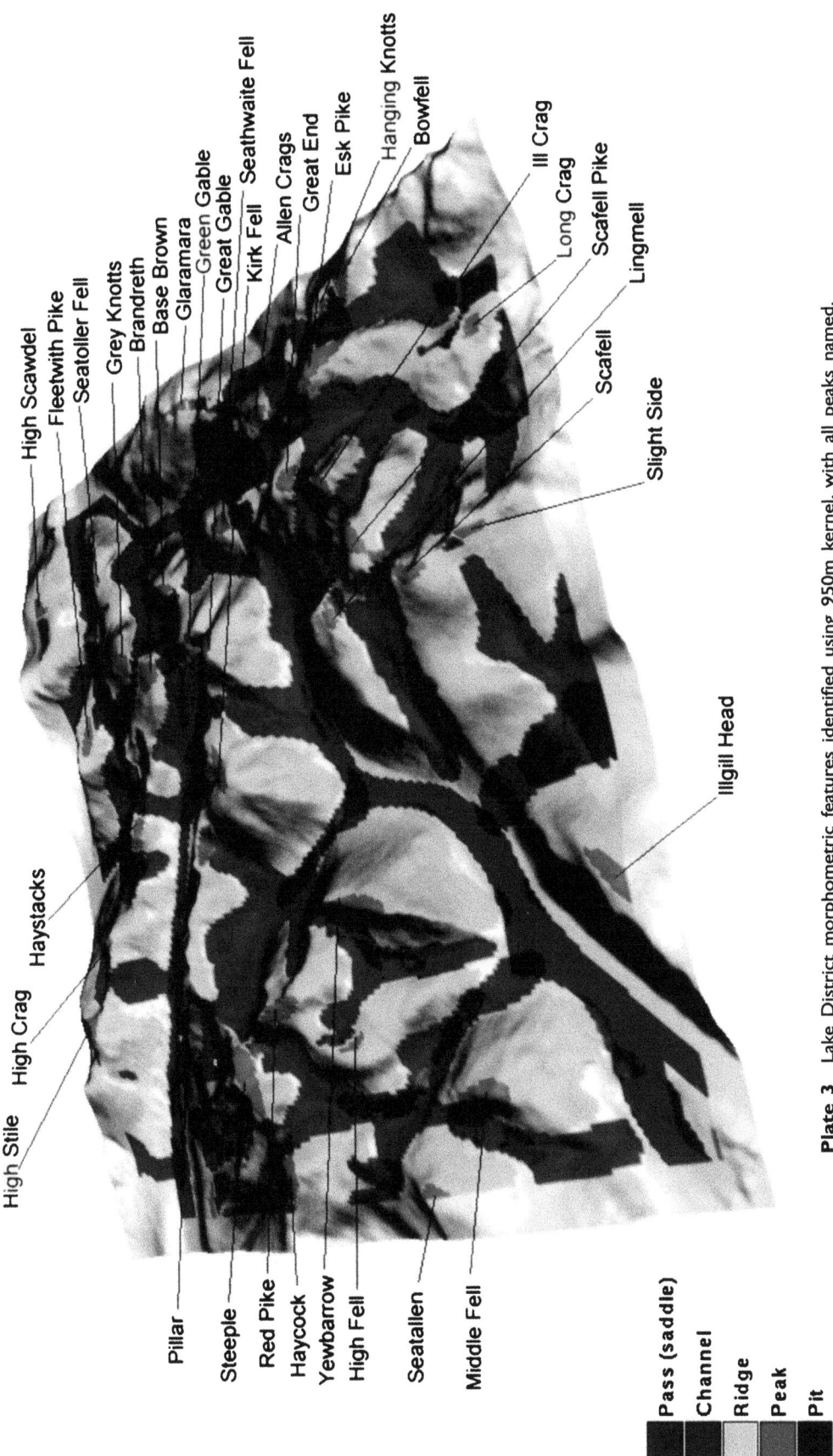

High Scawdel

High Stile

Haystacks

High Crag

Pillar

Steeple

Red Pike

Haycock

Yewbarrow

High Fell

Seatallen

Middle Fell

Fleetwith Pike

Seatoller Fell

Grey Knotts

Brandreth

Base Brown

Glaramara

Green Gable

Great Gable

Kirk Fell

Seathwaite Fell

Allen Crags

Great End

Esk Pike

Hanging Knotts

Bowfell

Ill Crag

Long Crag

Scafell Pike

Lingmell

Scafell

Slight Side

Illgill Head

Pass (saddle)

Channel

Ridge

Peak

Pit

Plate 3 Lake District morphometric features identified using 950m kernel, with all peaks named.

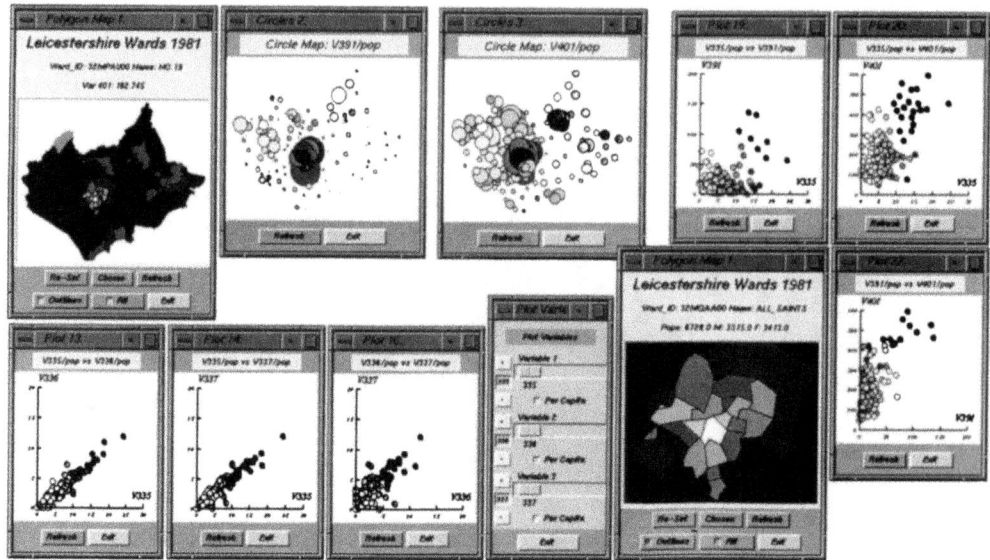

Plate 4 cdv – Cartographic visualisation for enumerated data.

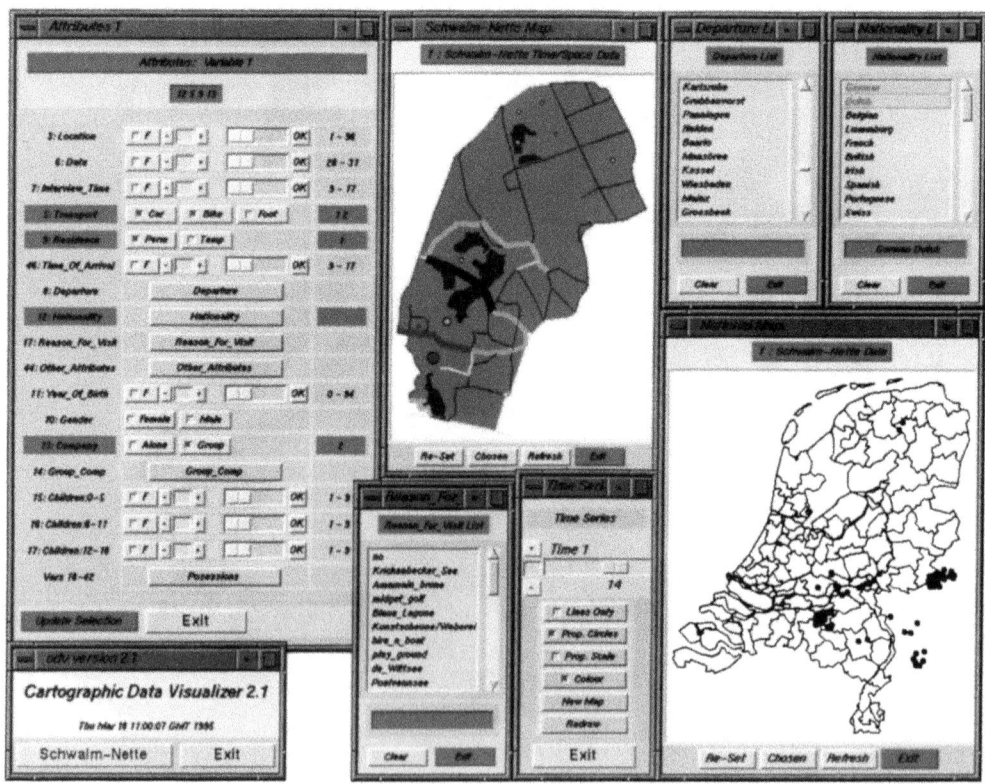

Plate 5 cdv – Cartographic visualisation of people in time and space.

The importance of determining efficient and reliable algorithms for lower dimensional topological primitives can be seen from the fact that the higher dimensional algorithms use the lower dimensional algorithms repeatedly.

Further work is required in this area to handle the limitations of co-ordinate resolution: the test described uses equality as its test condition. Inequality tests against tolerance values derived from co-ordinate resolution may give a suitable solution.

APPENDIX

Algorithm to determine intersection of an edge and a triangular face

We consider a triangular face defined by the co-ordinates of its three vertices. (A, B, C), and an edge (straight line segment) defined by the co-ordinates of its two endpoints (G, H). We assume that the triangle is non-redundant (i.e. A, B, and C are not collinear), and that none of the five points are coincident. We have used the notation:

xA: x co-ordinate of A, etc.

xBA: $xB - xA$, etc.

1. From A, B, C write down the implicit equation of the plane through the triangular face:

$$ax + by + cz + d = 0$$

where

$$a = yBA * zCA - zBA * yCA$$
$$b = zBA * xCA - xBA * zCA$$
$$c = xBA * xCA - xBA * xCA$$
$$d = -(xB * a + yB * b + zB * c) \qquad \text{(10.A.1)}$$

2. From G and H write down the parametric line equation of the edge:

$$x = xG + xHG * u$$
$$y = yG + yHG * u$$
$$z = zG + zHG * u \qquad \text{(10.A.2)}$$

For the whole line u takes all values from $-\infty$ to $+\infty$. For the edge (which is the segment between G and H), $0 \le u \le 1$.

3. Find the value of u at the intersection point between the plane and the line, and hence the co-ordinates of intersection point P:

$$u = \frac{-(a * xG + b * yG + c * zG + d)}{a * xHG + b * yHG + c * zHG}$$

if $0 \le u \le 1$ then intersection is within the line segment $\qquad \text{(10.A.3)}$

The line and plane are parallel if the denominator of the expression for u is zero. This case needs special treatment, which is not described here.

4. Do a point in triangle test for P in ABC. (We know that P, A, B, and C are coplanar.) To do this we use a parametric equation for the triangle ABC:

$$x = xA + xBA * s + xCA * t$$

and similarly for y and z.

For points in triangle ABC, $s \geq 0$, $t \geq 0$, $s + t \leq 1$

This can be written in matrix form:

$$\begin{pmatrix} x \\ y \\ z \end{pmatrix} = \begin{pmatrix} xBA & xCA \\ yBA & yCA \\ zBA & zCA \end{pmatrix} \begin{pmatrix} s \\ t \end{pmatrix} + \begin{pmatrix} xA \\ yA \\ zA \end{pmatrix} \qquad (10.A.4)$$

To find s, t co-ordinates of point P requires inversion of the matrix in the above expression. In general a $3 * 2$ matrix cannot be inverted. To allow inversion, we create an additional column. To ensure that this column is linearly independent of the two existing columns, we use the vector product of the first two columns.

$$\begin{pmatrix} xBA & xCA & yBA*zCA - zBA*yCA \\ yBA & yCA & zBA*xCA - xBA*zCA \\ zBA & zCA & xBA*yCA - yBA*xCA \end{pmatrix} \begin{pmatrix} s \\ t \\ 0 \end{pmatrix} = \begin{pmatrix} x \\ y \\ z \end{pmatrix} - \begin{pmatrix} xA \\ yA \\ zA \end{pmatrix}$$

Invert this to solve for s and t at point P:
writing

$$f = yBA * zCA - zBA * yCA$$
$$g = zBA * xCA - xBA * zCA$$
$$h = xBA * yCA - yBA * xCA$$
$$\det = f^2 + g^2 + h^2, \text{ then}$$

$$\begin{pmatrix} sP \\ tP \\ 0 \end{pmatrix} = \frac{1}{\det} \begin{pmatrix} yCA*h - zCA*g & zCA*f - xCA*h & xCA*g - yCA*f \\ zBA*g - yBA*h & xBA*h - zBA*f & yBA*f - xBA*g \\ f & g & h \end{pmatrix} \begin{pmatrix} xPA \\ yPA \\ zPA \end{pmatrix}$$

The third row is just expressing the fact that PA is linearly dependent on BA and CA.

The first two rows give a pre-inverse for the original matrix, so:

$$\begin{pmatrix} sP \\ tP \end{pmatrix} = \frac{1}{\det} \begin{pmatrix} yCA*h - zCA*g & zCA*f - xCA*h & xCA*g - yCA*f \\ zBA*g - yBA*h & xBA*h - zBA*f & yBA*f - xBA*g \end{pmatrix} \begin{pmatrix} xPA \\ yPA \\ zPA \end{pmatrix}$$

$$(10.A.5)$$

Test resulting values of s and t to see if they satisfy triangle conditions: $s \geq 0$, $t \geq 0$, $s + t \leq 1$.

REFERENCES

BOWYER, A. and WOODWARK, J. (1983) *A Programmer's Geometry*. London: Butterworths.

DGIWG (1994) *Digital Geographic Information Exchange Standard (DIGEST),* edn 1.2, January 1994. Digital Geographic Information Working Group (DGIWG).

MANTYLA, M. (1988) *An Introduction to Solid Modeling.* Rockville: Computer Science Press.

NEWMAN, W. M. and SPROULL, R. F. (1979) *Principles of Interactive Computer Graphics,* 2nd edn. New York: McGraw Hill.

PREPARATA, F. P. and SHAMOS, M. I. (1985) *Computational Geometry.* New York: Springer Verlag.

RIKKERS, R., MOLENAAR, M. and STUIVER, J. (1994) A query oriented implementation of a topologic data structure for 3-dimensional vector maps. *Internat. J. Geogr. Inform. Syst.* **8,** 261–270.

PART THREE

Spatial Analysis

CHAPTER ELEVEN

User-centred intelligent spatial analysis of point data

STAN OPENSHAW and TIM PERRÉE

11.1 INTRODUCTION

This chapter outlines the development of a new approach to spatial analysis in GIS environments. An attempt is made to develop an abstract but intuitively obvious spatial analysis process that is designed to be readily understandable to the typical non-statistical minded end-user of GIS. The precision and complexity of conventional spatial statistics and quantitative geography are replaced by a visual form of presentation that provides a simpler and more direct means of communicating and interpreting advanced spatial analysis technology. A prototype system is described, and assessed using synthetic data.

This chapter presents a new approach to spatial analysis that is designed to be relevant to the current data-rich GIS era. The first GIS revolution involved the *management* rather than the analysis of geographic information. Its very success has led to a geographic information explosion and created the need for a second GIS revolution; one that is focused upon the *analysis* and *use* of the geographical information that now exists. This requires the development of new analytical tools that are sufficiently powerful to make good use of the information riches and thus permit users to further capitalise on their investments in GIS technologies (Openshaw, 1994b,c). A very important need is to create new and more efficient ways of performing exploratory spatial data analysis (EDA) tasks relevant to GIS environments. According to Haining (1990) EDA involves the description of spatial data, the identification of statistical properties, and the preliminary identification of data structure, '. . . with the objective of encouraging hypothesis formulation from the data' (p. 4). Many users are probably more interested in packaging EDA as a function they can activate whenever they wish to explore spatial information for pattern or relationships. Unfortunately the spatial dimension presents many problems. Previous attempts to resolve these problems has produced spatial analysis methods that are too complex, perhaps too theoretically sophisticated, and usually too hard for many GIS users to cope with (Openshaw, 1991). As a result there have been few routine and successful applications of spatial analysis in GIS by end-users working outside of research environments.

Spatial data is inherently complex. While some significant patterns can be found using existing methods, many more undoubtedly remain undiscovered. The challenge of uncovering new patterns in spatial data is important, and its importance is emphasised by the large number of spatial datasets that now exist. Openshaw (1993) talks about the criminal waste of geographical information by both non-analysis and poor analysis of those key data that need to be properly analysed, either because it is in the public good to do so or because there are overwhelming commercial or other imperatives for its full and proper analysis. However, in seeking to use geographical information it is also important to ensure safe and appropriate use or else another type of GIS crime might be invoked (Openshaw, 1994a).

At one time it was thought likely that GIS systems would be able to service the complete spectrum of user needs and that the principal obstacle in a spatial analysis context was the coupling of existing analysis technology to proprietary GIS systems; see for example, Anselin *et al.* (1993). This now seems to be increasingly irrelevant as many of the end-user needs would appear to lie outside the areas for which existing, largely research focused, spatial analysis techniques exist. There is a strong argument that the safe and appropriate analysis of geographical information requires the development of new and easy to use tools based on new thinking, and new systems that are focused on the spatial analysis of geographic information in the context of the applied end-user. It is important to recognise that increasingly the end-users are no longer solely academic researchers or statistical experts, and they have a different set of needs. Neither spatial statistics nor quantitative geography has yet faced up to this particular challenge.

There are also some other major user-related problems that need to be addressed. Specifically, many users are not used to data riches, there is a lack of appropriate methods to help them exploit this new situation, and there is confusion about what analysis technology is needed. The novelty of the emerging new opportunities means that there is no pre-existing set of clearly identified user needs and requirements that can be used to specify computer systems. As Openshaw (1995a) argues, '. . . in many ways the potential applied spatial analysis opportunities that may be perceived by researchers to exist still need to be defined, demonstrated, and then converted by a programme of awareness raising into new sets of user needs and tools'. Maybe there is a need to shake both users and developers out of a highly complacent state so that they can start to become more aggressively adventurous in their exploitation of the new opportunities that GIS has provided them with, and to persuade them of a need to move on from where the more conventional methods stop. Maybe the key issue is to determine what users would want from spatial analysis if they actually knew what was now possible. One way of approaching this is to develop prototype systems with which end-users can explore new possibilities for analysis.

There are five principal issues, from the end-user's point of view, that need to be faced:

1. The apparent complexity of existing spatial analysis technology that often instils a neurosis of fear in the user about their own statistical inadequacies. For example, excellent books such as Haining (1990), Upton and Fingleton (1985), Griffith (1988), Cressie (1991), Anselin (1988), and Ripley (1981) are unintelligible and will remain unreadable to most GIS end-users.
2. The inherent inability of existing methods to provide useful results in many important practical applications.

3. The difficulty that end-users often experience in understanding what the results mean because of the nature of the statistical language that is commonly used to report them.
4. It is very difficult to use results that are not understood as the basis for decisions or actions.
5. Finally, there is a need to communicate the meaning and significance of the spatial analysis results to others.

These end-user concerns are essentially to be able to understand and trust the results of spatial analysis so that they can be used in managerial and decision making processes. This is quite different from the traditional research oriented focus of the spatial analyst or the quantitative geographer. A Type I Error probability has probably little relevance to many decision and policy making contexts. The decision maker wants a black or white answer not a qualified one; for instance, either there is a problem here or there is not. The fuzzy or qualified 'maybe' outcome does not help anyone as it implies uncertainty, suggests that the analyst does not know what is going on, and provides no information that end-users can cope with. This might be the correct outcome from an academic and research perspective but it is unhelpful to the policy or decision maker because it provides no basis for decisions. It is, therefore, no longer sufficient to 'sit on fences' but to try and meet the needs of the end-user more explicitly by developing appropriate technology that provides what they want in a form they can understand.

11.2 SOME GENERIC DESIGN PRINCIPLES

In attempting to develop an end-user oriented approach to spatial analysis some important issues emerge that question much of the conventional thinking that underpins current approaches. The principal need is to develop a style of spatial analysis that the users of GIS can use, feel comfortable with, and believe in. Experience suggests that these aspects cannot easily be added as an afterthought to methods that were created by experts for experts. Instead, end-user friendliness and ease of comprehension need to be built into both the methods and the software from the beginning. It is also striking that virtually every aspect of current spatial analysis technology is either seeking too much precision or attempting to be too complex and sophisticated. The analysis process is regarded as a science, based on rigor and as much precision as possible. Rigor is clearly an important virtue but it does not necessarily require the highest possible degree of precision regardless of the nature of the application. Consider, for example, mapping; the cartographic origins of GIS are still so dominant that it is usually considered that mapping has to be a highly precise process. This is an admirable goal for a national mapping agency but is it relevant for mapping data in a spatial analysis context in which a subtle change of class intervals can completely change the results? Conventional cartography is quite often too precise given the uncertain nature of the data it represents. It provides a level of accuracy that is spurious and in practice unnecessary. Often the end-user merely wishes to see results that matter and, probably nothing else.

There are two important design issues that need to be resolved: (1) the *form* of visualisation being used to communicate information to the end-user; and (2) the *content* of the information being communicated. Conventional cartography of raw

Table 11.1 Requirements and characteristics of this new approach to intelligent spatial analysis

1. The user should not need to know in advance precisely WHERE, WHEN, or WHAT to look for.
2. There is a need to incorporate knowledge, where it exists and when it is relevant.
3. It is unreasonable to expect users to possess many genuine *a priori* hypotheses that can be properly tested.
4. The tools have to be easy to use by non-experts.
5. The technology needs to be inherently safe so that the results are believable, appropriate to the context in which they were produced, and provide a basis for subsequent decisions and actions.
6. The results should be robust given the realities and nature of spatial information.
7. The technology should be flexible and comprehensive.
8. The analysis procedures should be capable of providing new insights and acting as an intelligent assistant working with, rather than against, the user in a creative partnership.
9. The methods of analysis have to be understandable to the end-users even if that understanding is at the level of plain English description.
10. The methods should deal with the principal needs for generic and application independent forms of spatial analysis.
11. The methods should resolve complex spatial data into simpler information about significant patterns or anomalies within the data.
12. The technology should be able to look within maps and be independent of study region boundaries.

or semi-processed data from a GIS may look colourful but it is usually inappropriate as a basis for analysis and communication. The acute observer may, with luck, good eye sight, and a sufficiently simple pattern, discover something of interest; but in practice this seldom happens! In spatial epidemiology the last pattern to be discovered by these means was over 150 years ago. GIS has done virtually nothing to improve this situation. Analysing spatial data by simply mapping it is a most inefficient, and wholly GIS inappropriate approach to the analysis of inherently complex spatial information. Colourful map wallpaper is for interior decorators not spatial analysts!

Deciding what to communicate is also fraught with traditions that can point in the wrong direction. It is hard to imagine a less appropriate approach to spatial analysis in a GIS context than that which is sometimes advocated by spatial statisticians and quantitative geographers. The statistical technology is too complex; it is highly elitist, being understandable only by a few; it is very partial in what it offers; it is highly fragile, with its validity often resting on naive and, sometimes, suspect assumptions about the nature of spatial data; and the quality of the results is almost totally dependent on the skills of the user. It ignores the nearly universal lack of a high level of statistical expertise amongst the end-users of GIS. The need is for user friendly systems designed for clarity of exposition rather than sophisticated systems of massive statistical complexity. Even when performed to a highly competent professional level, what end-user could use results produced by methods they may never be able to comprehend?

A new class of spatial analysis methods is needed. They should be exploratory rather than confirmatory because the latter serves to mislead. If inference is used it should merely be seen as a measure of performance that has no critical thresholds, other than perhaps the feeling that small probability values are better than large. It may also be necessary to allow for multiple testing, but without putting too much emphasis on the results or relying too heavily on testing null hypotheses that may be wholly inappropriate or wrongly specified for a GIS context. Computational intensive statistical methods particularly bootstrapping and Monte Carlo simulation can be used to try to ensure that only meaningful and valid results are passed to the user. Finally, it is important to avoid being too precise; most applications do not need it. There are real limits to what can be achieved by geographical analysis in a GIS context. The major requirements and characteristics of this new form of spatial analysis are outlined in Table 11.1.

11.3 SYSTEM DESIGN

Searching spatial databases for patterns and relationships is a common requirement of many spatial analysis procedures. This search process is often hard because of data complexities, data overload, multiple data domains, and lack of prior knowledge of what to look for and where to find it. In some systems the user controls the search in an interactive graphical environment; see for example, SPIDER, (Haslett *et al.*, 1990), REGARD (Unwin, 1992, 1994), and ISP (Nagel, 1994). This allows a user to understand aspects of the spatial structure of the data but its use is restricted to expert users and fairly simple low dimensional data sets. The other extreme is the fully automated and 'black box' approach to spatial analysis of Openshaw's Geographical Analysis Machine (GAM), Openshaw *et al.* (1987). In GAM there is no user interaction and the results are automatically generated. This approach has some attractions; particularly, its spatial comprehensiveness and independence of user skill level. However, there are also some problems; mainly related to the computationally intensive brute force search, its 'black box' nature, and the complexity of the underlying technology. Moreover, there is nothing to see other than a final and fixed set of results that may convey little or no understanding of the structure in the data or how the results were obtained. It would be helpful to develop a new approach that retained the best features of both visual interactive graphics and GAM but with none of the disadvantages.

This leads to the conjecture that the GIS end-user is more likely to trust, understand and use an intelligent spatial analysis tool that they can watch as it searches a GIS database for localised patterns in spatial data. The hope is that the end-user can, by observing an animation of the search process, develop a better understanding of what the results mean, and moreover is able to improve this understanding by watching the analysis process being repeated on library data sets that contain different types of known data patterns.

There are four components to this system:

1. An automatic exploratory mechanism that searches spatial data for map patterns under its own control.
2. A measure of search result performance that can be used to guide the search.
3. An interpretation module that statistically processes the results to highlight the important information, taking into account data uncertainty and multiple testing effects.
4. A means of visualising the search process via animation, emphasising the significant patterns and allowing user interaction with them.

Table 11.2 Principal design assumptions relating to the system

1. Separation of visualisation of the spatial analysis process from the spatial analysis process to allow for heterogeneous computing
2. No real time interaction with the analysis process
3. No user interaction with the animation creation process
4. Computational methods used to ensure high quality and consistent results are produced
5. Built-in self-checking mechanisms to identify inconsequential results
6. Based on generic principles
7. Capable of being upgraded in a modular fashion as the constituent technologies improve

These components need to be complementary. The search process has to be efficient in its coverage of geographic space and has to provide results that help the user understand the nature of any spatial structure being highlighted. In the described system the visualisation takes place only after the analysis is complete. This separation of analysis from visualisation allows the use of highly complex analysis methods without being restricted to the need for instant results. This approach enables the analysis process to be performed on a remote machine, before the end-user interacts with the results of the analysis. The visualisation system is highly interactive, but uses results generated earlier. There is also no assumption that the user is an expert able to interact with a complex statistical analysis process or a visualisation package that may involve a learning curve at least as lengthy as a GIS system. The skill level of the end-user should not constrain the analysis process, but equally nor should the interactive visualisation of analysis results put limits on what the analysis process can do.

The system design assumptions are contained in Table 11.2, and the overall system structure is given in Figure 11.1.

Although the final visualisation is a map of some kind, the map is not a traditional one, containing vast amounts of detail that impresses the bystander and oppresses the user. Detail that cannot be used is little more than map junk. Only the critical information should be presented and then in a highly simplified graphic format. If you ask the question, 'Where are the cancer clusters?', it should create a display of icons to broadly represent where they are located and their general distribution. There is no need for any further precision. The results are 'real' and significant if the underlying analysis technology is explicitly designed to ensure that this is so. Statistics that only experts understand, and then perhaps do not believe, are simply not relevant here. If the results are worth noting then they should be presented to the user in a form that can be clearly understood. It is an important function of the statistical processing that black and white decisions are made which are likely to be correct. The end-user does not want to know what the likelihood of being right or wrong is. Such knowledge is a distraction, it is seldom understood, and it only confuses. Of course, this runs contrary to conventional academic opinion. It is not that detailed statistics are of no use, it is merely that the statistical processing system should be clever enough to take them into account before forming an opinion that the end-user can simply understand, trust, and use or reject. Whether the resulting system can be trusted is a matter for objective evaluation, but dodging the issues by dumping a heap of largely incomprehensible statistics on the end-user, and then declaring that interpretation is their responsibility, helps nobody. So a simple iconic presentation of the results that matter is important in developing

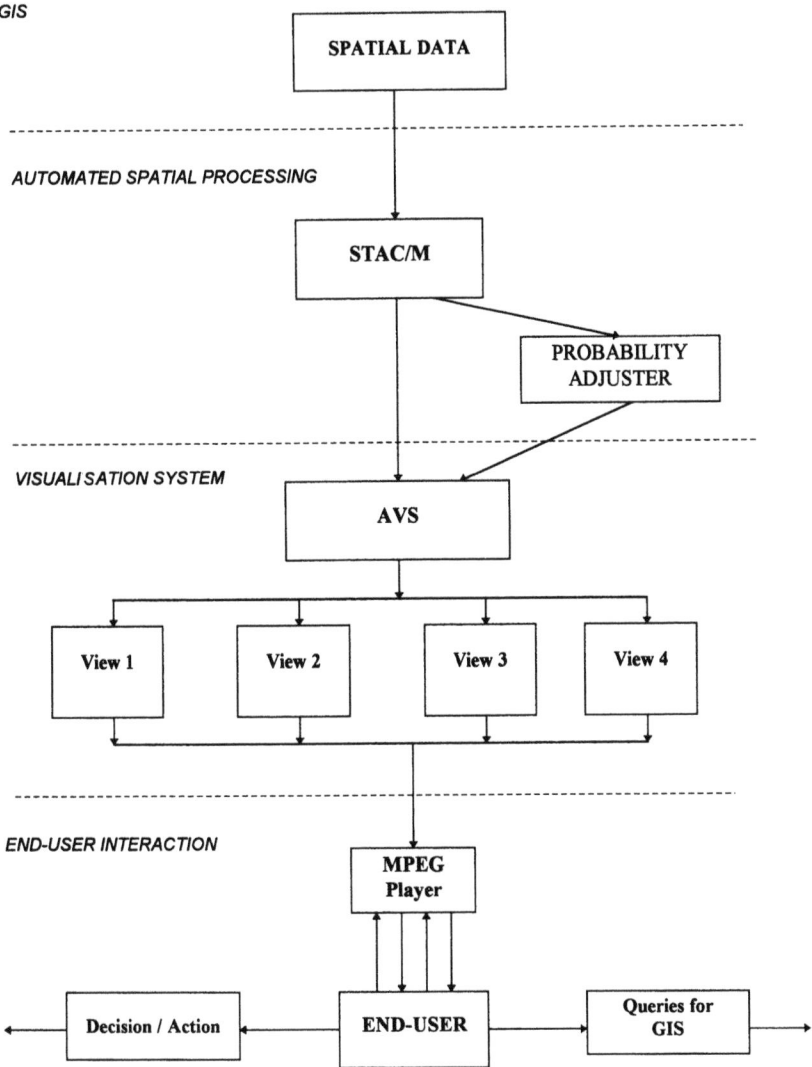

Figure 11.1 System structure.

an end-user oriented approach to spatial analysis. It is designed to be geographically fuzzy rather than precise and to be abstract rather than detailed, because that is what the communication process seems to require.

11.4 BUILDING A SMART POINT DATA ANALYSIS SYSTEM

11.4.1 The spatial search and analysis module

A useful spatial analysis method able to detect clusters in point referenced disease data is the Kth nearest neighbour method of Besag and Newell (1991). This method is a version of the Geographical Analysis Machine that uses a search based on observed cases. It was developed in the context of detecting what are termed 'data anomalies' in childhood cancer databases but is more generally applicable as a means of analysing point data for evidence of localised clustering. With this method a small positive integer K is chosen and is used to draw circles centred on each case of the disease in turn and passing through the location of its $(K-1)$th nearest neighbour. In each of the circles the number of individuals at risk is determined and if this number is unusually low (i.e. there are an unusually large number of cases per population at risk, so that resulting Poisson probability is sufficiently small) the circle is plotted on a map of the region. The plotted circles represent the anomalous areas, and a range of maps may be produced using different values of K. The results for any specific K value can be compared because the method takes into account the effect of variations in the underlying population distribution. This simple method suffers from a number of problems. In particular, the spatial search is incomplete as it only looks in regions around existing cases, which implies a highly localised cause and may well fail to adequately represent the locations and extent of any anomalies it finds; it ignores multiple testing with a tendency for a high false positive rate; and it cannot detect small clusters because of the loss of a degree of freedom, which probably only matters most in the analysis of very rare diseases. It is also a statistical method that most GIS end-users may find hard to use or understand.

Openshaw (1994d, 1995b) describes the development of what is termed a Space Time Attribute Creature (STAC), an artificial creature that is created to search for patterns in GIS databases under the control of a Genetic Algorithm (GA). This is used here to develop a much simplified version that implements the Besag–Newell Kth nearest neighbour method. The search is determined by three parameters; X and Y values for the circle centre, and a value for K. The search strategy no longer depends on the location of existing cases but can examine any location, to virtually any level of spatial precision, within the study area. This approach makes fewer arbitrary prior assumptions about the data, which may affect the results. The program that contains this GA implementation is called STAC/M, the Space–Time–Attribute Creature/Movie.

The STAC/M uses a basic genetic algorithm to drive the exploratory search. The three parameters are encoded as bit strings. The X, Y co-ordinates are each stored as 14 bit numbers and the value for K coded in 8 bits. This is sufficient precision for current purposes; giving the X, Y co-ordinates a 10 m resolution and allowing values of K in the range from 0 to 256. The basic genetic algorithm operates as follows:

Step 1. Generate a population of PSIZE random X, Y, K bit strings.

Step 2. Evaluate each location using a Poisson probability (P) to represent the fitness of any given X, Y, K triplet. The function $1 - P$ is used.

Step 3. Create some children bit strings by applying genetic operators to the data. This occurs as follows: with probability of pc randomly select two parents according to fitness, apply a two point crossover of their bit strings, apply mutation (probability pm), and inversion (with probability pi). Only above average performing bit strings are subjected to crossover and there is an offspring restriction to avoid premature loss of diversity; see Goldberg (1989), Davis (1991) for details.

Step 4. Evaluate the performance of the M new bit strings.

Step 5. Randomly select M worst performing members from the original PSIZE strings and replace.

Step 6. Repeat steps 3 to 5, a number of times; each constitutes one generation.

Here the GA parameter PSIZE is set at 64, pc is 0.95, pm is 0.01, and pi is 0.02.

GAs are a well established and highly effective optimisation procedures able to rapidly seek out the optimal values of very complex functions. Here the GA is being used to find X, Y, and K values that have the highest probability of being the location of data anomalies; in effect it is seeking to minimise the Type I error probability. Note that there is no explicit test of hypothesis as the Poisson probability is merely being used in a descriptive sense; as a measure of performance that happens to be normalised on the range 0.0 to 1.0. The intention is to use the search process generated in step 1 and step 5 to visualise the spatial analysis function being performed. The version of the GA used here was developed to be gradual rather than rapid, ensuring that the developing search can be watched, and to be informative about the locations of possible multiple different anomalies. The GA searches for a global optimum result, but in the current context it would be more useful if it also located other significant but sub-optimal results. To this end, the GA crossover process was modified to include a spatial neighbourhood effect on the second parent selection process. The search for other sets of pattern can be achieved by removing the data that contributed to the first set of results and then re-running the STAC/M analysis on what is left.

The results for each GA generation are output to a series of ASCII files containing details of each bit string; the location, the search radius, the Poisson probability of the observed number of cases being due to chance; the number of cases within the search radius, and the corresponding population at risk. These files form the base data sets used in the subsequent visualisation of the spatial analysis process.

11.4.2 The statistical processing system

The STAC/M results only really constitute a data screening. The GA driven search process is highly effective in finding localised data anomalies, but some of the seemingly interesting results could well be due to small number effects or data unreliability. One way of handling this problem is to compute a more robust measure of the fitness function used in the search process. A simple and effective modification is to replace the simple Poisson probability calculation by a bootstrapped version. The objective is to avoid extreme results that might mislead the GA due to data uncertainty. The bootstrap operates as follows. For a particular X, Y, K combination retrieve the data that lies within the implicit circle focused on location X, Y that has K cases within it. Re-sample this data subset to generate 2000 bootstrap samples of the same size, note that the sampling is done with replacement. Now select and use as a measure of fitness the median Poisson probability value. This avoids extremely small probabilities being generated due to small number effects which are highly variable.

Another problem is that of multiple testing. The GA selects the best result by testing multiple hypotheses and this might also produce misleading results; see Hochberg and Tamhane (1987). Geographers have long been guilty of ignoring the multiple testing problem; see for example, Getis and Ord (1992). Indeed, virtually every map generated that presents the results of statistical testing suffers from multiple testing problems; the larger the number of zones the greater the problem and the more misleading the extreme results may become. Once a result is declared significant then many will believe it is real and even attempt to explain it in terms of possible causes. Therefore it is important to correct the results for multiple testing and not mislead the end-user who may well be quite unaware of the problem.

There are a number of potentially suitable methods for correcting for multiple testing. Recently, Benjamini and Hochberg (1995) describe a modified sequential rejective method that is used here. The method works as follows:

Step 1. Assume the testing of N null hypotheses H_1, H_2 $\cdots$ H_N are based on the probability values $P_1, P_2, \dots P_N$. They are sorted into ascending order.

Step 2. Let k be the largest i for which P_i is less than or equal to (i/N)alpha where alpha is the false discovery rate, which can be regarded as equivalent to the usual Type I error significance level.

Step 3. Reject all H_i $i = 1, 2 \cdots k$

The number of hypotheses being tested is the total number of unique X, Y, K values generated during the search process. The GA search is so effective that when clustered data are analysed the probabilities become extremely small and survive the adjustment procedure. This adjustment procedure can be improved by re-sampling so that the distribution characteristics as well as the extreme values of the distribution of results can be handled; see Westfall and Young (1993). The principal problem is the two to three orders of magnitude increase in compute times so this aspect is left for future development on a faster high performance computing platform.

11.4.3   The visualisation and animation system

Previous computer movies of mapped data have tended to focus on animation as the principal means of analysis. Dorling and Openshaw (1992) and Openshaw et al. (1994) describe how to animate space–time data series that have been subject to two- and three-dimensional space–time smoothing. Here the objective is quite different. The purpose is to provide a meaningful visualisation of the spatial analysis process itself. The visualisation is designed to: (1) inform the end-user about the nature of the spatial analysis that has been performed, (2) offer insight into the structure of the results by watching their emergence, and (3) highlight areas of particular interest. These visualisations operate on both the raw and multiple testing adjusted data and are designed to offer additional information about the analysis process over and above that provided by the statistical aspects. For example, the geographical distribution and location of data abnormalities as they appear during the search process is itself of considerable interest. Both the trail of extreme results scattered over the map and those locations where the results seem to concentrate are offering very different insights. Some results will reflect the GA search process and how it interacts with the underlying data. Others may provide a meta-analysis of the results that transcends that offered by any of the statistics; for example, the locations of heaps of anomalies compared to much more spread out and even distribution is indicative of two very different types of spatial patterning. The challenge here is to develop the visualisation process so as to emphasise these potentially insightful and pattern reflective aspects so that an intuitive understanding can be developed. They should show the interactions between a spatial analysis tool and the data being analysed. It is argued that this concept can be generalised and extended into higher dimensional space. The aim here is to make visible not the statistical aspects of spatial analysis but the intuitive, the artistic, and the creative. There is no wish to test hypotheses or state results with a precise but probably misleading, and certainly difficult to comprehend, level of quantitative precision. Instead the aim is to provide visualisations of significant information about patterns and process that may be of interest, that would appear to be unusual in relation to reference benchmarks, and which encapsulate intuitive notions regarding the feel of the spatial data under analysis. The hypothesis is that the viewer may gain more understanding of the structure within the spatial data set by visualising the progress of the whole search process, than by only visualising the final set of results.

The UNIX-based Application Visualisation System (AVS) is used here because this state-of-the-art scientific visualisation software offers a flexible range of visualisation options, including an animation capability. AVS is one of the most commonly used visualisation software packages. It is designed to be easily extended through user-written modules, and many additional modules are freely available. AVS is also well suited to the rapid prototyping of different visualisation methods – the so-called 'plug, play and throw away!' approach to development. However, the need for an end-user orientation precludes letting the user have access to AVS. That would violate the design objective of ease of use. Instead the plan was to offer the user a series of prepared views of the spatial analysis process that emphasised different aspects, and use AVS purely as a means of generating the animations.

Perrée (1994) describes some early experimentation with AVS, using various visualisation approaches to present animated views which attempted to maximise understanding of the STAC/M results. The initial work focused on using circles against a background map of the study area. The location of the circles corresponded to the location of the STAC/M X,Y values and the circle radius and colour represented attributes such as the size of the search area and the probability of the result. This initial visualisation method mapped the performance measures to the circle colour using a range of colours from blue to red. The red circles (i.e. the hottest) represented the best search locations and the blue circles (i.e. the coldest) represent the worst. The K values represents a distance value and this is mapped to the circle radius. The resulting visualisation gave the first views of the STAC/M data as a series of various coloured circles against a background map.

These initial studies revealed the need for some improvements in the visualisation of the STAC/M results. As a result of considerable experimentation, a number of visualisations have been developed, each of which emphasises different aspects of the spatial analysis process. See Figure 11.2 for examples of the different views.

View 1. The X,Y locations of STAC/M bit strings are mapped as white circles of uniform size. This view helps the end-user to focus on the STAC/M search process itself without the distraction of additional information on bit string performance and search radii.

View 2. The STAC/M bit strings are mapped as coloured circles. The performance of the bit strings is represented by the colour; red for the best and blue for the worst. The circle diameter represents the geographical search radius. This view focuses on where the search process is taking place.

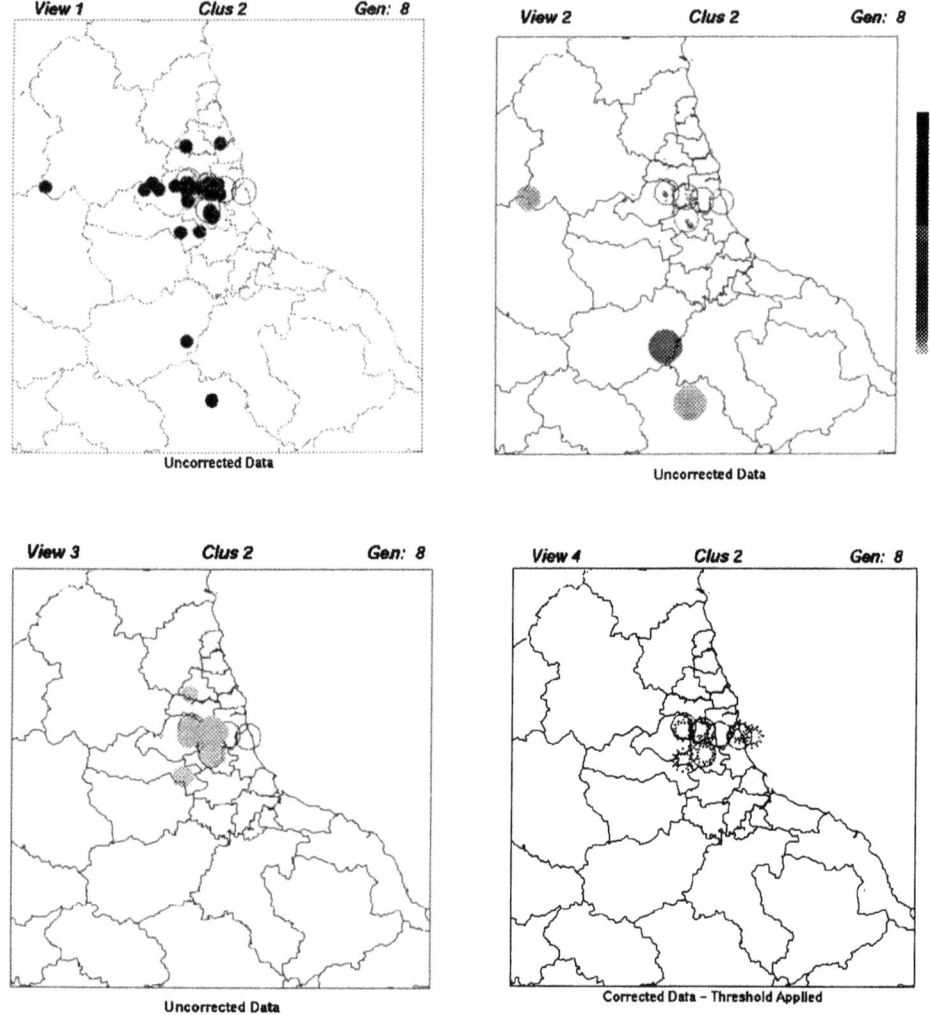

Figure 11.2 Four standard views showing different aspects of the STAC/M search results.

View 3. The bit strings are mapped as red circles, with the circle size corresponding to performance. A non-linear mapping, performance raised to the power of 4, is used to emphasise the location of the best performers.

View 4. The best performing STAC/M bit strings are represented using a star icon. The stars represent the locations of the most significant results. The stars do not overlap each other, and each star may represent one or several high performance bit strings. This iconic representation emphasises where the best performing STAC/M bit strings are located.

The AVS system is used to create these views and to save them as MPEG format movies that the user can play back later on a variety of different hardware platforms. The animation of the results is an important feature of this approach to intelligent spatial analysis; a sequence of still images on paper cannot convey the results in the way that an animated movie sequence can. The user's task is to watch and identify whether any of the results appear surprising or interesting. The avail-

Table 11.3 Summary of results

Number of evaluations	Data 1	Data 2	Data 3	Data 4
	648	1510	1445	1378
Number of uncorrected $P < 0.01$ at generation				
(View 1) 5	1	6	1	1
10	42	12	1	1
15	64	20	1	1
20	64	24	1	1
25	64	24	1	1
Number of corrected $P < 0.01$ at generation				
(View 2) 5	0	4	0	0
10	41	4	0	0
15	64	9	0	0
20	64	8	0	0
25	64	15	0	0
Number of uncorrected $P < 0.05$ at generation				
(View 3) 5	1	6	1	1
10	43	12	1	1
15	64	20	1	1
20	64	24	1	1
25	64	24	1	1
Number of corrected $P < 0.05$ at generation				
(View 4) 5	0	5	0	0
10	42	11	0	0
15	64	19	0	0
20	64	23	0	0
25	64	23	0	0

ability of the same standard views for different data sets, including purely synthetic data with known results, allows users to train themselves and to compare real data analysis against library benchmarks. This is useful both to improve their analytical performance and to gain confidence that the technology works.

11.5 RESULTS

This new method is evaluated using synthetic data containing known spatial structure. Four different data sets that simulate a rare disease with a frequency similar to that of childhood leukaemia are used. The data consist of a random component with a small systematic clustered part. The clustering process is represented by a two-dimensional gaussian distribution. Data 1 consists of a single big cluster focused at a single site. The method homes straight in on it. It is so obvious that all four views pick it up. The iconic display clearly and unambiguously identifies it. Data 2 is more difficult in that there are a number of much weaker clusters, located fairly close together. The question now is whether or not the methods find one or some or all of them. The task is harder but the iconic display identifies all the clusters. Data 3 is even more difficult in that the clustering is weaker and located in areas that might be expected to be even more difficult to find. Indeed, the iconic display fails to detect any clustering. However, all the other views are giving indications of fairly weak clustering in the correct areas of the map. Data 4 is purely random and the iconic display suggests nothing. The other views either show nothing at all or pick up weak clustering around the edges of the map where the search process has identified locations with very weak clustering due to edge effects. This illustrates the effectiveness of the genetic search process. These unintended data artefacts are detected as such. These results look very different and are much weaker than the results reported for Data 3.

The results for the four different data sets are contained in the series of MPEG format movies that were created using the four standard views. These MPEG movies are publicly available via the Centre for Computational Geography's Home Page on the World Wide Web at http://www.geog.leeds.ac.uk/research/ccg.html. They are also summarised in Table 11.3.

In data sets 1 to 3, there is a competition between the genetic algorithm's search for extreme results and the need to quickly find ever more extreme results to withstand the effects of the multiple testing adjustments. The cost of this correction is that the sensitivity of the method is reduced and the weak cluster in Data 3 is lost. However, this is probably for the best since in practice cluster detection causes considerable public anxiety so being conservative is a very useful attribute.

11.6 CONCLUSIONS

The chapter has outlined a new approach to spatial analysis that is regarded as being of some general utility to GIS. It has argued that current spatial analysis methods are often too complex, too assumption dependent, too precise, and too narrowly focused to be of much practical use to most of the potential end-user community. A new approach is advocated that seeks to combine the benefits of sophisticated analysis methods with the visual appeal of interactive map graphics to devise a simpler and more end-user friendly technology. This different style of approach seeks to combine intelligent automated methods of analysis, that use computational methods to ensure robustness, with a very simple visual presentation of the results based on an iconic interpretation of what they mean. The user is not expected to be, or to become, a statistician, a quantitative geographer or a computer scientist. Instead the technology has been designed to be almost instantly understandable in an intuitively meaningful way. This new approach is designed to support vague types of spatial analysis question; such as 'What should I know about what is happening?', and 'Whereabouts should I be looking?'. The system responds by offering equally abstract responses; such as 'It's over here', 'It's here, there and everywhere', 'It's very rare'; and so on. The precise map details and data queries that contribute to the detected patterns and relationships, together with the underlying statistics and data, still exist. It's just that they have been abstracted out and removed from view.

The question of how believable the results might be still needs to be answered. The curious can always look more closely at the details that supported the conclusions being presented. However, the technology is designed to be intrinsically safe, so that the user could simply trust the results that are presented. This might seem at first sight to be a recipe for disaster because it is counter to current thinking about how to apply spatial statistical methods. However, it can be no worse than not having analysis technology or having to rely on experts as the interface, most of whom are quite unable to communicate what the results mean to end-users, because they are not the end-users. It is far better to build intelligent software that can answer the questions that interest users and at the same time yield reliable results. The most the user should be asked is to specify their required level of certainty that the results are significant, so that only results above this threshold are displayed. This is purely a matter of deciding how risk-averse to be.

The immediate objective here is to specify, in a general way, a spatial analysis technology that operates in an abstract manner and produces understandable high quality results that users can cope with. This would consist of a readily understandable interface to some of the most complex spatial analysis technology available. A longer term task is to demonstrate that users can develop and rely on an intuitive, soft, qualitative, impressionistic understanding of spatial analysis that transcends the standard approaches. It is argued that the general paradigm described here provides a basis for developing a new generation of user friendly, intelligent and appropriate spatial analysis tools relevant to data rich GIS environments.

REFERENCES

ANSELIN, L. (1988) *Statistical Inference for Spatial Processes*. Cambridge: Cambridge University Press.

ANSELIN, L., DODSON, R. F. and HUDAK, S. (1993) Linking GIS and spatial data analysis in practice. *Geogr. Syst.* 1, 3–23.

BENJAMINI, J. and HOCHBERG, Y. (1995) Controlling the false discovery rate: a practical and powerful approach to multiple testing. *J. Roy. Stat. Soc.* B, **57**, 289–300.

BESAG, J. and NEWELL, J. (1991) The detection of clusters in rare disease. *J. Roy. Stat. Soc.* A, **154**, 143–156.

CRESSIE, N. A. C. (1991) *Statistics for Spatial Data*. Chichester: Wiley.

DAVIS, L. (1991) *Handbook of Genetic Algorithms*. New York: Van Nostrand Reinhold.

DORLING, D. and OPENSHAW, S. (1992) Using computer animation to visualise space–time patterns. *Environ. Plan.* B **19**, 639–650.

GETIS, A. and ORD, J. K. (1992) The analysis of spatial association by distance statistics. *Geogr. Anal.* **24**, 189–206.

GOLDBERG, D. E. (1989) *Genetic Algorithms in Search, Optimization and Machine Learning*. Reading, MA: Addison-Wesley.

GRIFFITH, D. (1988) *Advanced Spatial Statistics*. Dordrecht: Kluwer.

HAINING, R. (1990) *Spatial Data Analysis in the Social and Environmental Sciences*. Cambridge: Cambridge University Press.

HASLETT, J., WILLS, G. and UNWIN, A. (1990) SPIDER – an interactive statistical tool for the analysis of spatially distributed data. *Intern. J. GIS* **4**, 285–296.

HOCHBERG, Y. and TAMHANE, A. C. (1987) *Multiple Comparison Procedures*. New York: Wiley.

NAGEL, M. (1994) Interactive analysis of spatial data. In DIRSCHEDL, P. and OSTERMANN, R. eds, *Computational Statistics*, Heidelberg: Physica, pp. 295–314.

OPENSHAW, S. (1991) Developing appropriate spatial analysis methods for GIS. In MAGUIRE, D. J., GOODCHILD, M. F. and RHIND, D. W, eds, *Geographical Information Systems: Methodology and Practical Applications*, vol 1, London: Longmans.

OPENSHAW, S. (1993) GIS Crime and GIS criminality. *Environ. Plan.* A, **25**, 451–458.

OPENSHAW, S. (1994a) GIS Crime and Spatial Analysis. In Proceedings of GIS and Public Policy Conference, pp. 22–35, Ulster Business School.

OPENSHAW, S. (1994b) Computational human geography: towards a research agenda. *Environ. Plan.* A, **26**, 499–505.

OPENSHAW, S. (1994c) A concepts rich approach to spatial analysis, theory generation, and scientific discovery in GIS using massively parallel computation. In WORBOYS, M. F. ed., *Innovations in GIS*, London: Taylor & Francis, pp. 123–138.

OPENSHAW, S. (1994d) Two exploratory space–time attribute pattern analysers relevant to GIS. In FOTHERINGHAM, S. and ROGERSON, P. eds, *GIS and Spatial Analysis*, London: Taylor & Francis, pp. 83–104.

OPENSHAW, S. (1995a) Developing intelligent and user friendly spatial analysis tools for data rich GIS environments. Proceedings of Joint European Conference and Exhibition on Geographical Information. The Hague. Vol. 2, 417–25.

OPENSHAW, S. (1995b) Developing automated and smart spatial pattern exploration tools for geographical information systems applications. *Statistician*, **44**, 3–16.

OPENSHAW, S. and PERRÉE. T. (1995) Intelligent spatial analysis of point data. Proc. 50th Session of the International Statistical Institute, Beijing, China, August 1995.

OPENSHAW, S., CHARLTON, M., WYMER, C. and CRAFT, A. (1987) A Mark 1 Geographical Analysis Machine for the automated analysis of point data sets. *Intern. J. Geogr. Inform. Syst.* **1**, 335–358.

OPENSHAW, S., WAUGH, D. and CROSS, A. (1994) Some ideas about the use of map animation as a spatial analysis tool. In HEARNSHAW, H.M. and UNWIN, D. eds, *Visualisation in GIS*, Chichester: Wiley, pp. 131–138.

PERRÉE, T. (1994) The investigation of a scientific visualization and animation method for viewing the results and evolution of a cluster detecting genetic algorithm – a new approach to exploratory spatial analysis. Unpublished dissertation, School of Geography, University of Leeds.

RIPLEY, B. D. (1981) *Spatial Statistics.* Chichester: Wiley.

UNWIN, A. R. (1992) How interactive graphics will revolutionise statistical practice. *Statistician* **41**, 347–351.

UNWIN, A. R. (1994) REGARDing geographical data. In DIRSCHEDL, P. and OSTERMANN, R. eds, *Computational Statistics*, Heidelberg: Physica, pp. 315–326.

UPTON, G. J. and FINGLETON, B. (1985) *Spatial Statistics by Example*, vol 1: *Point Pattern and Quantitative Data*. New York: Wiley.

WESTFALL, P. H. and YOUNG, S. S. (1993) *Resampling Based Multiple Testing*. New York: Wiley.

CHAPTER TWELVE

Developing an exploratory spatial analysis system in XLisp-Stat

CHRIS BRUNSDON and MARTIN CHARLTON

12.1 INTRODUCTION

Since the very early days of commercial GIS, there have been calls for the incorporation of spatial analysis functionality into existing software systems. It is argued that although GIS software provides powerful tools for the management and manipulation of spatial data, there is little provision for statistical modelling and investigation of relationships between the different items of spatial data (Goodchild, 1987; Fotheringham and Rogerson, 1993). Various attempts have been made to meet these demands, using a varied range of approaches. At a very basic level, it is possible to write information from a GIS into a file, and read this into a statistical package such as SAS or SPSS, carrying out the analysis in this package. Any relevant information may then be written out of the statistical package and read back by the GIS.

A major difficulty of this approachis that a great deal of computing time is spent transferring data from the internal format of one package into a text format (for an export file) and then converting this into the internal format of another package! To counter this difficulty, some attempts have been made to read and write to and from the GIS's internal data storage memory directly using the statistical package (Kehris, 1990).

There is, however, a major drawback with either of these approaches – the data analysis that may be carried out is essentially non-spatial. Most of the techniques offered by statistical packages analyse only the attribute information relating to the geographical objects stored in the GIS – any data relating to the absolute or relative spatial location of the objects is discarded. When dealing with geographical data it is often the case that this information is of the greatest concern.

Part of this problem can be solved if a the link is made directly to a computer program in FORTRAN or C. In this way, it would be possible to implement analytical techniques relevant to spatial problems. Another similar approach is that of Arc/S-Plus, where Arc/Info is linked to the S-Plus package. S-Plus, unlike other statistical packages, is designed primarily as a programming environment. An example of its use is given by Diggle and Rowlingson's SPLANCS package (Rowlingson and Diggle, 1991; Gatrell and Rowlingson, 1993).

Of all of the methods suggested above, the last one seems to offer the most for researchers in spatial analysis. The S-Plus environment provides a test-bed for the development of new spatial analysis techniques, using a language that is well-suited to statistical techniques. Since S-Plus uses a command line approach to programming, rather than the compile-link-run cycle of C or FORTRAN, it allows a more interactive approach to algorithm development. The ease with which algorithms may be developed is further enhanced by the matrix and vector handling capabilities of S-Plus.

However, this approach still has some shortcomings. First, the linkage between Arc/Info and S-Plus is hardly seamless. Mapping operations are carried out in Arc/Info, and statistical operations are carried out in S-Plus. The way data are stored and manipulated in the two packages are very different, as are the syntactical structures of their command languages. A more consistent 'look and feel' for the GIS and spatial analysis operations would be desirable.

The previous objection could be dismissed as cosmetic – but there are further setbacks. Many spatial analysis techniques currently under development are exploratory in nature – and often involve the use of interactive graphical techniques (Getis, 1993). In particular, they may require interactive cartographic representation of geographical data (Haslett *et al.*, 1990) and linked multiple views of the data. Thus, developing new exploratory techniques may require the creation and control of maps directly from the algorithms created. Without complete coupling such control is cumbersome.

Finally, there is still the problem that the information exchanged between the two packages is restricted to attribute data. There are some spatial techniques, such as the fractal based methods of Batty and Longley, where data relating to the geographical entities themselves are required. A vector or list-based programming language such as S-Plus would be ideal to process lists of vertices such as coastlines, or roads, which could be interpreted as fractals, if only there were a simple way of transferring this information.

All of this suggests that some form of completely linked spatial analysis system would be of use – but the question that remains is 'how may this be achieved?'. The above discussion suggests that linking two distinct packages is not a good approach. It would be better if the spatial analysis functionality were incorporated into the GIS. However, this is not an easy task. For example, although AMLs provide some very useful facilities for the automated control of Arc/Info, it would be very difficult to carry out many of the operations required for spatial analysis using this language. For example, it would be virtually impossible to implement Diggle's K-function estimation procedures relying only on AML. Generally, most GIS macro languages lack the flexibility to carry out much of this type of analysis, as they lack features such as matrix manipulation, or a broad range of list processing operations.

In this chapter, an alternative approach is proposed. Instead of extending the analytical capabilities of a GIS, why not extend the geographical data handling and mapping facilities of a package designed for statistical programming? In this way, if the package has reasonably sophisticated graphical capabilities, and is able to handle geographical data structures, it is possible to build mapping facilities which may be seamlessly addressed and controlled within the language of the package. Here, an attempt is made is made to provide such a facility within Luke Tierney's XLisp-Stat package (Tierney, 1991). This public domain package, based on the Lisp programming language (Moyne, 1991), has properties making it well suited for such a task. In the following sections, XLisp-Stat will be described briefly, and the implementation of an interactive spatial data handling and mapping system in this package will be discussed.

12.2 THE LISP LANGUAGE

LISP is primarily designed as a LISt Processing tool. A list differs from a simple array in several ways. Firstly, it does not have a pre-determined length. An array's dimension must be declared at the beginning of a program, whereas a list's length may be an initially unknown quantity. Indeed, the length of a list will often vary as a program runs its course. Secondly, a list's elements do not all have to be the same data type. A list consisting of the elements 31, 'March' and 1995 would be allowed, where the second element is a character string, and the others are integers.

Another important feature of lists in Lisp is that they can contain other lists as elements. This recursive structure makes it possible for lists to represent data structures such as binary trees. It is also possible for a list to have no elements.

A very unusual feature of Lisp is that programs are themselves lists. The elements of a list can be any kind of symbol, and there is no reason why symbols cannot represent actions to be carried out by Lisp. This feature of Lisp can be very useful. For example, if creating a procedure to calibrate a regression model, then a y-variable transformation function could be passed as a variable.

An important consequence of the equivalence of data and programs in Lisp is that object orientation is relatively easy to achieve. An object is generally perceived as an entity containing both information (data) and methods (functions). Since data and functions are the same in Lisp, an object may be represented as a list of data lists and function lists. This rather natural representation for objects in Lisp has led to most modern Lisp systems having strong support for object oriented software development.

12.2.1 The XLisp-Stat package

Recently, it has been observed that Lisp is not only useful as a programming language for artificial intelligence, but may also be useful for statistical processing. To this end Tierney's book describes the Lisp-Stat system, which is essentially a Lisp interpreter with extra statistical functionality. XLisp-Stat is an implementation of the Lisp-Stat system, which has versions for MS Windows, Apple Macintosh and Amiga personal computers as well as an X-Windows version for UNIX and VMS-based workstations, building on Betz' XLisp-Plus system.

Again, lists are very useful here. For example, the data for a two sample t-test could be thought of as two lists of measurements, and a general one-way ANOVA could be thought of several lists of measurements – one for each treatment. In the latter case, a list of lists of measurement may also be a useful data representation.

In addition to the computational extensions to Lisp, there are also graphical extensions. These include features which are specific to exploratory statistical analysis. Particularly relevant are the 'multiple view', brushing and linked plot facilities. Given a multivariate data set (with more than two or three variables), it is difficult to visualise all of the information for each case simultaneously. However, it may be possible to show the information using several two- or three-dimensional scatter plots or histograms. The information lost in doing this is the association between each of the variables. It cannot be seen which cases in one plot relate to the cases in another.

However, this type of question can be answered using 'brushing' or linked view techniques. Using the mouse, if the above plots are linked, then by highlighting part of the histogram with a mouse, the corresponding points on the scatter plot will also be highlighted. Alternatively, if points on the scatter plot are selected, then corresponding parts of the histogram will be highlighted. Brushing is similar to this, but instead of requiring a point-and-click selection techniques, the mouse is attached to a 'moving window' so that as it moves over a graph, points within this window are continuously selected and deselected. This is also the principle on which the SPIDER (Haslett *et al.*, 1990) package is based.

XLisp-Stat not only provides graphical features of the type discussed here, but also makes use of Lisp's object oriented features. Linkable graphs and histograms are objects, and in particular they have the property of 'inheritance'. Inheritance is a characteristic feature of object oriented programming, where new objects are defined in terms of existing ones, but with some modifications. As suggested previously, objects have 'methods' – which are algorithms associated with the entities they represent, and 'slots' which are associated data items. For example, a graph window object may have a method, called 'draw-axes', to draw its own axes, and slots for the minimum and maximum values for the scales of these axes. A new object could then be defined, inheriting from the graph. For example, a log–log plot would inherit many features of a graph window object, including all of the window-related abilities, but would have a different axis-drawing and point-plotting methods, to allow for logarithmic, rather than linear, scaling.

This approach leads to very efficient coding. When objects are very similar to their 'ancestors' only unique features need to be coded, the rest being automatically acquired by the inheritance mechanism. It also makes objects very easy to control. In the above example, both the original graph window and the new logarithmic version would have methods called 'draw-axes', although one had a slightly different algorithm to the other. If another algorithm, designed to produced multiple plots, were to send both objects a 'draw-axes' message, they would each carry out their appropriate method, without the sender of the message needing to determine precisely their object type. In fact, the controlling code could have been written before the logarithmic graph object was even thought of – it simply requires that any object it attempts to control has some method called 'draw-axes'. In this way, inherited objects are implicitly fitted into the existing system, allowing apparently seamless extensions of XLisp-Stat's capabilities without modification of the existing operating environment. In XLisp-Stat, an initial set of objects provides linkable graph windows and several windowing interface control objects, such as slider bars, buttons and tick boxes. In the next section, an object-oriented approach exploiting these will be used to provide exploratory spatial data analysis facilities.

12.3 GEOGRAPHICAL DATA HANDLING IN XLISP-STAT

Having introduced XLisp-Stat as a Lisp-based, object-oriented *statistical* analysis package, it must now be considered how it may be expanded to become a package capable of *geographical* analysis. One helpful step along this road is the ability to handle complex numbers. XLisp-Stat can carry out mathematical operations (such as addition, multiplication or taking logarithms) on complex numbers with little more difficulty than using real numbers. Complex numbers are useful for the representation of geographical information since they can be thought of as representing points in a two-dimensional plane (sometimes referred to as the Argand plane, or the C-plane. Thus, a single value can represent a point in a plane, rather than the two values that would be required in a system only capable of handling real-valued numbers. This also makes computation of quantities like distance easier. This is simply the magnitude of the difference between two complex numbers. The use of complex numbers adds to the brevity of algorithmic notation already achieved by using a list-based language.

12.3.1 Points, lines, areas and attributes

In vector based geographical information systems, the basic geographical entities are points, lines and areas. In this section it will be demonstrated how these may all be represented in Lisp. The previous section suggests that spatial data should be modelled using complex numbers. First, a point coverage may be thought of as a list of complex numbers. This simplistic representation already allows a variety of spatial manipulation techniques to be applied. Line coverages may be similarly represented. A line in a vector-based GIS is represented by a list of vertices, which are simply points which may be joined together, 'join-the-dots' style. A single line can therefore be represented by a list of complex numbers. A line coverage, which consists of several lines, can be represented by a list of lines; that is a list of lists of complex numbers. Again, spatial manipulation of lines in relatively simple. Applying a distance function (as defined earlier) to each sequential pair of vertices in a line and summing these gives the length of a line.

Finally, consider how areas may be represented in Lisp. A simple model (although not the most storage efficient) would be to store an area as its boundary in the line form discussed above. In this case the line would be 'closed off', so that its first element was joined to its last, in order to enclose a two dimensional region. Again, a list of such lists allows an entire coverage of polygons to be represented. Clearly, this approach also allows functions such as boundary length (or perimeter) of zones to be computed in the same way as suggested for line data. The only modification required in the case of a length computation would be to add the distance between the first and last vertices (complex numbers) in the list.

Thus, all basic geographical entities may be represented as either a list of complex numbers, or a list of lists of complex numbers. However, this only covers the representation of the geographical objects themselves, and does not consider any attributes that may be attached to them. This could form the basis of a desktop mapping system, but not a complete GIS. The next stage is to consider the addition of attributes to the data model. This may be achieved by introducing a second list, complementing the geographical entity, whose elements would correspond to the elements of the geographical list, but would contain the associated attributes of each entity. Thus, a list of county perimeters would by accompanied by a list of, say, unemployment figures. Giving both of these data items to a map drawing function would then produce a county-based unemployment choropleth map.

The flexibility of the list as a data structure allows this idea to be extended. Instead of storing a single attribute value for each element in the list, it would be possible to store an entire list of values. This provides several opportunities. For example, instead of storing a single value, such as an unemployment rate, a list of values could be stored – for example, several census-based variables for each county. There is also no restriction that the attribute attached to each geographical entity has to have a fixed length. It would be reasonable for each area-based zone to have a list of its adjacent zone indices attached to it, or for each road in a network to have a list of indices of other roads with which it has a junction. In this way topological information can also be stored as attribute lists.

12.3.2   Algorithms

A few examples in the previous section suggest that many spatial data handling algorithms can be implemented in very compact code. In this section, some more examples of coding will be considered, demonstrating firstly some basic data handling procedures, and then some spatial statistical methods.

A point-in-polygon algorithm is often useful, in order to determine which points in a data set lie within a polygon, or to count the number of point events (for example incidents of some disease) lying within a set of areas (for example Census ward boundaries). One method of determining whether a point is in a polygon is the 'angle-sum' method. Suppose it is intended to determine whether a point X is within a polygon P. Then this method takes the angles t_i subtended between Z and each pair of successive vertices in P and adds these (Figure 12.1). If the point lies within the polygon, the angle will be an odd multiple of 360 degrees, but if it lies outside, it will be an even multiple.

The function may be made more efficient if a check is initially made as to whether Z lies in the bounding rectangle of P. This may then be extended, using the *map-elements* function, to be applied to an entire list of points, to test whether each one is within a polygon. Another extension would also allow a single point to be tested for containment in a number of polygons. A variation on this is used to define a *count-elements* function which is given a list of polygons and a list of points, and returns the number of points in each polygon as an attribute list. Using techniques similar to these, a series of spatial data handling functions have been defined (Table

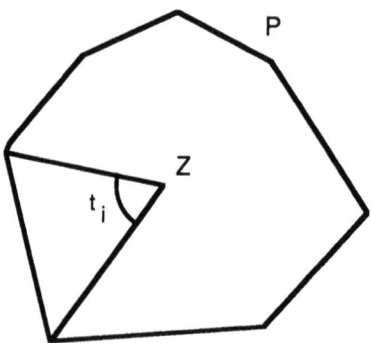

Figure 12.1   The angle sum point-in-polygon algorithm.

Table 12.1 A set of geographical data handling functions for XLisp-Stat

Function name	Action of function
Within	Tests whether a point is within a polygon
Count-items	Counts number of points in a list of polygons
Sum-items	Sums attributes of a list of points over their containing polygons
Contiguity	Returns contiguity for a list of polygons, by testing for common boundaries
Bounding box	Returns two points, the bounding box for a point, line or area coverage
Centroids	Returns a list of point centroids for a list of polygons
Areas	Returns a list of areas given a list of polygons
Perimeters	Returns a list of perimeters given a list of polygons (or lines)

12.1). These are intended to provide a useful grounding for the manipulation of spatial data in XLisp-Stat.

Note that in the above list of functions, it is possible to determine a contiguity list. Such information, stored either in list form or as a contiguity matrix, is essential for several spatial analytical techniques. In XLisp-Stat it is possible to store this data in either form. The choice of which is most appropriate is usually dependent on computational efficiency. The sparser a contiguity is, the greater the savings that may be made in terms of storage and computational efficiency. An elementary example of spatial analysis using contiguity lists is that of spatial smoothing. A mean smooth (see e.g. Fotheringham, 1993) is an operation in which a value for each given zone in an area coverage is replaced by the mean value of itself and its adjacent neighbours. If there is a contiguity list for the neighbours of a particular zone in a list regions of zones, then the *select* function will pick the relevant zones out of this list, so that their mean may be computed.

Similarly, replacing the mean function with a median function would allow a median-smooth operation to be defined. A final development might be to create an arbitrary smoothing function, which took a function as a variable and applied this to the lists of contiguous attributes. This would provide scope for experimentation. For example, if a standard deviation function were used as a 'smoother', an index of local variability would be obtained. In addition to these more descriptive methods, it is possible to implement more formal inferential techniques.

12.3.3 Interactive maps in XLisp-Stat

It has hopefully been demonstrated that Lisp is an appropriate language for handling spatial information, but this would be of very little use without the ability to create maps. Fortunately, in XLisp-Stat there are several interactive graphics tools that allow maps to be created. These are implemented as objects in Lisp, and the inheritance feature of the Lisp object system is useful here to provide map objects based on these. There is an object in XLisp-Stat, 'Graph-Window-Proto', which may be used as the base object for several types of graphics window. This is a special type of object, referred to as a prototype (hence its name). Prototype objects are able to create clones of themselves, when sent the message 'new'. Thus, *graph-window-proto* is used to create new graphics windows. Its slots include information about graph scaling, whether it preserves aspect ratios if the window is re-sized, the colour, style and thickness of lines as they are drawn and several other items of data relevant to scale drawing of data. These items of information are all important when using a window of this type to display maps. The methods associated with the graph window include the ability to draw lines, points and filled areas (again important for mapping) and to add textual annotation.

A central method for *graph-window-proto* is the *re-draw* method. Whenever a window has to re-draw itself, perhaps because it has been eclipsed by another window, or iconised, or re-sized, this method is automatically called. The definition of the re-draw method is essentially the algorithm to create the graphic in the window. Recalling that one notion behind inheritance is that if a new object is created based on an older one, all of the methods of the old are contained in the new, and also that it is possible to overwrite some of the old methods, it is often the case that newly-defined graphics objects will overwrite the re-draw method. In this case, a map window object is created (called *map-window-proto*) based on *graph-window-proto*.

The contents of a map window can be classified according to the data entities they are visualisations of. Map windows can contain information based on either point, line, or area coverages. However, there are several ways in which each type of coverage may be visualised. In this map drawing system, a particular visualisation of a coverage is referred to as a *layer*. A layer can be thought of as a drawing on an overhead projector transparency, so that a map composition could contain several of these layers. In areas where two pieces of information were displayed on separate layers, the information that is visible would correspond to the uppermost layer of the two. A map-window-proto would therefore have some new slots added, a most important one being *layer-list* a list of the layers (ordered with the uppermost item last) that comprise the map composition visible in the window. A typical map window is shown in Figure 12.2. A list of the map layers provided is given in Table 12.2.

The map layers will themselves be objects. One reason for this is that each layer may then have an *exhibit* method. This, if given a window object, would cause the layer to be drawn in that window. Exhibit would be defined differently for each type of layer, allowing for different data representation techniques. However, in the re-draw method for the map window, it would only be necessary to send each item in the layer list an exhibit message, with the correct exhibit algorithm being chosen automatically. If a user wished to define a new type of map layer, then provided it had an exhibit method, it would be handled by the re-draw method in map-window-proto without modification. The fact that some map layer objects are already defined encourages this, since new map layers can inherit properties from the initial ones. This therefore allows users to experiment with, and extend the functionality of the map drawing facilities. Here, an integrated system is advantageous: although

Table 12.2 Map layer objects provided

Map layer name	Description
Powder layer	Small, coloured points representing a point coverage
Spot layer	Proportional circles, with size based on attributes, representing a point coverage
Line layer	Lines, of varying width and colour, representing a line coverage
Poly layer	Polygons filled with colours representing area attributes – choropleth mapping
Text layer	Annotation in text form

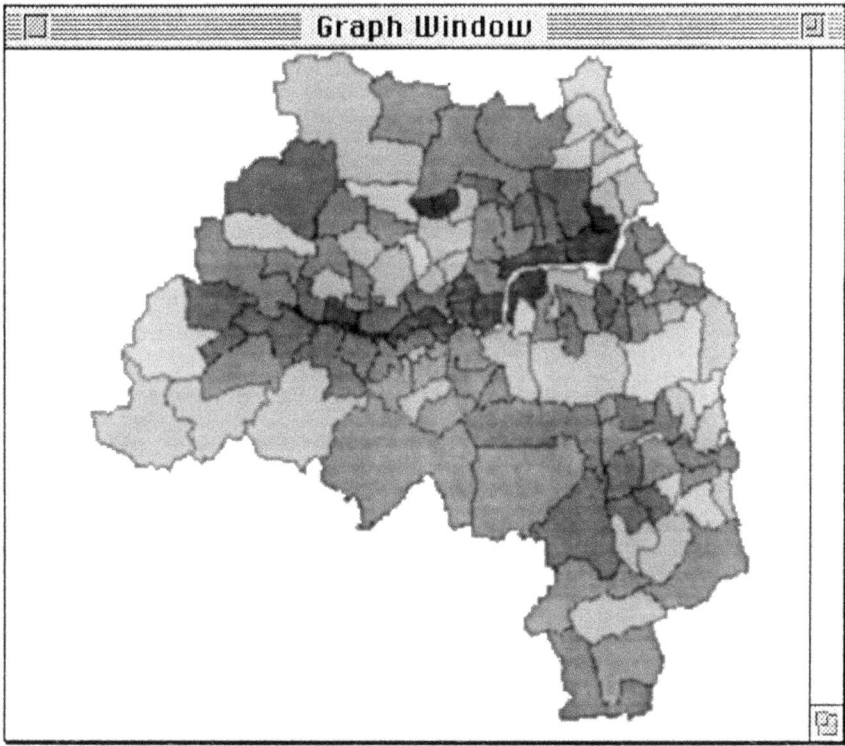

Figure 12.2 An example of a map window object in XLisp-Stat.

S-Plus also has object-oriented features, it would not be able to treat maps created using Arc/Info as S-Plus objects, and create new objects based on these.

A final, but important property of the map layers is that they may become part of the linked view system described in section 12.2. That is, if a histogram or scatter plot is used to represent some attribute data, then this could be linked to the map so that selecting points on the scatter plot would highlight the corresponding points, lines or areas in the map. This is currently achieved by giving each of the map layer objects a *selection* message. When this is sent, with an index referring to the particular point, line or polygon that is selected on a linked plot, it causes the colour of the geographical entity to change to the *selection* colour. This will default to yellow. The next time a selection message is sent, the object will revert to its original colour. Due to inheritance, the map window proto has a message called *set selection* which is sent to it by any linked graphical window at appropriate times, and this is re-defined here to send selection messages to map layers.

Finally, the interface of these map layers to the user of the package must be considered. Although the message-sending object system of Lisp is relatively easy to program, a simpler interface is required to allow efficient operation when spatial data is being explored interactively. To this end, a single function, with several optional parameters, called *map* has been created. This function has been defined so that it will create a map window, and add layers to it according to the data it is provided with. If it is given a polygon coverage, and a set of attributes, it will create a choropleth map. If it is given a line attributes coverage, it will create a line map. If it is given a point coverage and a set of attributes it will create a proportional circle map. The function returns a map-window object. If it is desired to add extra layers to this map, there are functions to create map layer objects called *line-layer*, *powder-layer*, *spot-layer*, *poly-layer* and *text-layer*. Used in conjunction with the *add-layer* message in the map window object, more complex map compositions may be created.

12.4 CONCLUSIONS

What has been proposed here is not intended to be a full-blown geographical information system – for example, it has no facility to help users digitise maps – but is intended to provide most of the GIS functionality need to carry out spatial analysis. The use of the Lisp language and complex numbers to process spatial data has demonstrated that at least those spatial data handling functions required for such tasks can be added to XLisp-Stat with relative ease. In particular, the object oriented extensions, and the graph window objects provided a simple means of adding cartographic features to the system. Due to the inheritance feature very little coding was needed, since most of the graphical and windowing capabilities of a map window were already provided in the parent object.

The object orientation of the system should also be useful for future users of the system, as the map objects created may themselves be the basis for other objects. For example, a map window capable of drawing world maps in several different projections could inherit from the map prototype, but then have functionality for computing planimetric coordinates for the different projections incorporated. The advantage here is clearly in complete coupling. The mapping and data processing features come from the same system, and may be freely mixed when developing new ideas.

To conclude this chapter, two final advantages of this system will be considered. First, the Lisp code used to create this extension to XLisp-Stat is portable with respect to all of the hardware platforms mentioned in section 12.2. The code was initially developed on a Power Macintosh, but has also been tested on a 66MHz 80486 processor based PC compatible platform and a Sun SparcStation running the OpenWin X-Windows based desktop environment. Whereas the Sun workstation was clearly the most powerful operating environment of the three, the system ran acceptably on the two personal computers. Since Macintoshes and IBM PC compatibles account for a very great proportion of all personal computers, it is hoped that this will make this system available to a broad range of potential users.

Another very important consideration is that XLisp-Stat, and several extensions written for it, are in the public domain. The XLisp-Stat package for all of the above platforms may be obtained from the FTP site umnstat.stat.umn,edu, where there is also a large library of Lisp code available. In particular, it is intended that the mapping code discussed here will also be available from this site in the near future.

REFERENCES

DIGGLE, P. J. (1983) *Statistical Analysis of Point Patterns*. London: Academic Press.
FOTHERINGHAM, A. S. (1993) Scale-independent spatial analysis. In GOODCHILD, M. and GOPAL, S. eds, *The Accuracy of Spatial Databases*. London: Taylor & Francis, pp. 221–228.

FOTHERINGHAM, A. S. and ROGERSON, P. A. (1992) GIS and spatial analytical problems. *Intern. J. Geogr. Inform. Syst.* **7**, 3–19.

GATRELL, A. and ROWLINGSON, B. (1993) Spatial point process modelling in a GIS environment. In FOTHERINGHAM, A. S. and ROGERSON, P. A. eds, *Spatial Analysis and GIS.* Taylor & Francis, pp. 147–163.

GETIS, A. (1993) Spatial dependence and heterogeneity and proximal databases. In FOTHERINGHAM, A. S. and ROGERSON, P. A. eds, *Spatial Analysis and GIS.* London: Taylor & Francis, pp. 147–163.

GOODCHILD, M. (1987) A spatial analytical perspective on geographical information systems. *Internat. J. Geogr. Inform. Syst.* **1**, 327–334.

HASLETT, J., WILLS, G. and UNWIN, A. (1990) SPIDER – An interactive statistical tools for the analysis of spatially distributed data. *Internat. J. Geogr. Inform. Syst.* **4**, 285–296.

KEHRIS, E. (1990) A geographical modelling environment built around Arc/Info. Regional Research Laboratory Initiative Research Report 13, Lancaster University.

MORAN, P. A. P. (1950) Notes on continuous stochastic phenomena. *Biometrika*, **37**, 17–23.

MOYNE, J. A. (1991) *Lisp: a First Language for Computing.* New York: Van Nostrand Reinhold.

RIPLEY, B. D. (1981) *Spatial Statistics.* New York: Wiley.

ROWLINGSON, B. and DIGGLE, P. J. (1991) SPLANCS: spatial point pattern analysis code in S-Plus. *Computers Geosci.* **19**, 627–655.

TIERNEY, L. (1991) *LISP-STAT: an Object-Oriented Environment for Statistical Computing and Dynamic Graphics.* New York: Wiley.

PART FOUR

Visualisation

CHAPTER THIRTEEN

Exploring urban development dynamics through visualisation and animation

MICHAEL BATTY and DAVID HOWES

13.1 INTRODUCTION

In this chapter, we show how the increasing availability of digital data at the micro urban level and software based on GIS and remote sensing are now essential in generating and testing new theories of urban form and dynamics. Here our concern is with assembling such data, exploring them in a preliminary way, while setting up methods for enabling appropriate visualisation and computer animation of urban space–time patterns. We illustrate these ideas with property parcel data for the Buffalo metropolitan region, first suggesting that the evolution of this region illustrates a surprising stability, consistent with theories of spatial self-organisation, then showing how basic problems of visualisation need to be explored across different spatial and temporal scales and through consistent methods for animating the data.

13.2 NEW IDEAS ABOUT URBAN DYNAMICS

Explaining urban development as an aggregate system of land use and economic activities has been the predominant theoretical approach to understanding and fore-casting the structure of cities for the last 50 years. This has led to models which are essentially static but it is increasingly clear that these types of theory are limited in their abilities to deal with the kinds of disequilibrium that all cities manifest. They are unable to accommodate spontaneous growth, or any surprising changes that depart from observed past trends, indeed any form of novelty or innovation. Contemporary urban theory, however, takes a different approach being largely based on the new non-linear dynamics which treats urban change as discontinuous in the manner of catastrophe and chaos. Yet although there are many new theories in the making (Wilson, 1981; Dendrinos, 1992; Allen, 1994), it has been difficult to make progress in empirically testing such theories which require much richer dynamic and more micro-spatial data than available or demanded hitherto.

Connected to these developments is a new concern for urban form. This builds on the idea that form, like dynamics, is random and irregular in the way it is generated locally. Substantial insights into these processes have emerged over the last decade, particularly in complexity theory and in the study of chaos. Chaotic systems are those in which very slight changes in initial conditions can result in manifestly different structures within a short time. Models of the way local decisions give rise to more global spatial structures are in fact at the essence of fractal theory in which processes exist across many scales. Some of these are already being applied to cities through spatial diffusion (Batty and Longley, 1994) and cellular automata theory (White and Engelen, 1993; Batty and Xie, 1994). Linking form to process has also been the recent concern of the weak chaos theory pioneered by Bak and Chen (1991) in their theory of self-organised criticality which seems particularly suited to explaining the evolution of cities.

The fact that we can begin to make any progress whatsoever with these types of theory is due to the emergence of new digital data sources and software for their handling. Data for many US cities pertaining to the location and attributes of all taxable properties is now available as standard GIS coverages, and physical data pertaining to natural features, demographic data and so on can be related consistently to this. New procedures for visualising very large data sets are available through GIS and remote sensing (RS) software, while methods for animating space–time data are being developed. So far, most of the applications of GIS to urban problems have been to mapping and data integration, and to ways of visualising the results of spatial analysis. However, the prospect now exists of using this technology as a basic tool in developing better urban theory, through new ways of visualising and animating data.

Here we will outline the rudiments of the data and information which we see as necessary for our somewhat grander theoretical interests. We will apply our analysis to the Buffalo metropolitan region, a city with around 1.3 million population in 1990 located in the Great Lakes where New York State touches the Canadian province of Ontario. Since the 1960s, Buffalo has made a rapid transition economically from an old, monocentrically structured industrial city to one which now contains almost no manufacturing but is based on a spatially diffuse mix of local and regional services. Our thesis here is that as Buffalo is making this transition, at some stage its urban form may change dramatically and our data and theory should be able to show this. To anticipate this, we are able to show that for the last 70 years, the city has been in a kind of stable equilibrium consistent with Bak and Chen's (1991) idea of a critical state, but as yet there is no sense of an impending transition. We are fortunate in having an excellent database at the micro-spatial level based on taxation data coded with respect to the age of the property: a kind of time series which represents the 'fossilisation' of the city as of now.

We will begin by sketching the evolution of the Buffalo region providing a traditional ecological view, but to explore these ideas further, we require a much more detailed picture. First we examine the problem of aggregating the data in space and time simply to show how necessary this is to visualise the data in any meaningful way. We then take this visualisation further by overlaying other physical data – first from digital elevation modelling and then from raw line feature data. This improves our ability to explain the pattern of urban development as well as in checking for consistency and error but it is largely a prelude to our animation. We then show how the animator can be constructed, arguing that such techniques are essential in thinking about development change, both backwards as well as forwards in time. This is work in progress. We conclude by outlining the next steps.

13.3   THE EVOLUTION OF URBAN FORM

Cities prior to the industrial revolution were either overgrown villages or concentrations of government and related functions. Their role was to service the agricultural economy. The internal division of functions within such cities was rudimentary, based on a well-defined core but with craft industry and residential activities mixed in a profusion of types around this core. Industrialisation changed all this with cities becoming much more spatially articulate. In the heyday of this era, the typical city consisted of a core based on commerce, an inner zone around the core comprising manufacturing industries, all set within some concentric organisation of residential activities, ordered according to the ability of different groups in the population to transport themselves to the core for work and to bid for residential space. The way the city grew appeared to be in waves of development emanating from the core, diffusing to the periphery.

This is a picture of cities in equilibrium but it is one that depends entirely on the structure of industrial society and that, we know, is passing. It no longer provides a good analogy to explain 'edge cities', or 'world cities' for that matter. There is nothing in the model that indicates how major transitions in urban form might emerge, nor is there any explicit acknowledgement of the important role of historical accident in the evolution of cities (Krugman, 1994). Buffalo provides an excellent case study for this theoretical dilemma. It is one of the clearest examples of an industrial city which has lost its original economic base and whose rationale is now based on health and educational activities with some minor functions pertaining to its proximity to Toronto. The notion of a monocentric city is no longer appropriate but there is little sense as to how the city might develop in the next 50 years. The safe answer would be to argue that it will simply spread out further and that older buildings, when vacated, will simply be demolished or abandoned. The truth is likely to be very different and will require the new kinds of theory whose data we are assembling and exploring here.

The data consist of 336 334 parcels, comprising all taxable properties in Erie County, collected annually by the State Board of Equalization and Assessment. Each parcel is coded by its x-y State Plane co-ordinates with a variety of attributes: two measures of land/property value but also information on type of use – residential or commercial/industrial, width and depth of lot, and most important for our purposes, the year of construction of the structure(s) on each parcel. The implied time series provides a picture of how the present city has been formed from the past, not how the past city has led to the present. This is important for the data cannot be construed as representing the actual evolution of present-day Buffalo although much of what exists now must represent structures which have been rebuilt on previous sites.

There is no way of testing our assumption that the series implied in the current data provides a reasonable representation of the way the city has evolved without assembling independent data from the past. We are attempting this in the same way that Kirtland *et al.* (1994) have morphed together the history of urban growth in the Bay Area but we are not able to report this yet. Examining the data in comparison with the growth of population in Erie County since 1830, we have some confidence that the demolition and rebuilding of structures has been quite regular during this period with a half life of around 48 years, but we must take on trust that this data can be generalised to a true time series.

There are, however, other problems in the data set; for example, 85 879 properties have not been coded with respect to age of construction. We are in the process of estimating the year of construction for this missing data using regression models and remote sensing which appear promising but we will not report these here (Batty and Howes, 1995). The major problems we are concerned with involve the visualisation of such large data sets. Using a 100 m grid on which to code the data (the GRID module in *ARC-INFO*), we show development at six 30-year time frames in Figure 13.1. Our sequence begins in 1820 although the first points in our data set go back to 1750. It is clear that the industrial city does not really get going until 1850 and only from 1880 is its dominant presence in the larger region apparent. Figure 13.1 shows the change in the number of parcels through time and this suggests that we think of this application as one of changes in urban development, not the growth of populations.

From Figure 13.1 and our more casual knowledge of Buffalo, the picture is one of a sparsely settled and turbulent frontier, often at war with the British in Canada until around 1820. Things settled after the Napoleonic wars. Until the mid-nineteenth century, the village of Buffalo grew and agricultural settlement stabilised. The city itself emerged between 1850 and 1880 with growth rates peaking in the period from 1880 to around 1920. Once the automobile era began, the city spread in

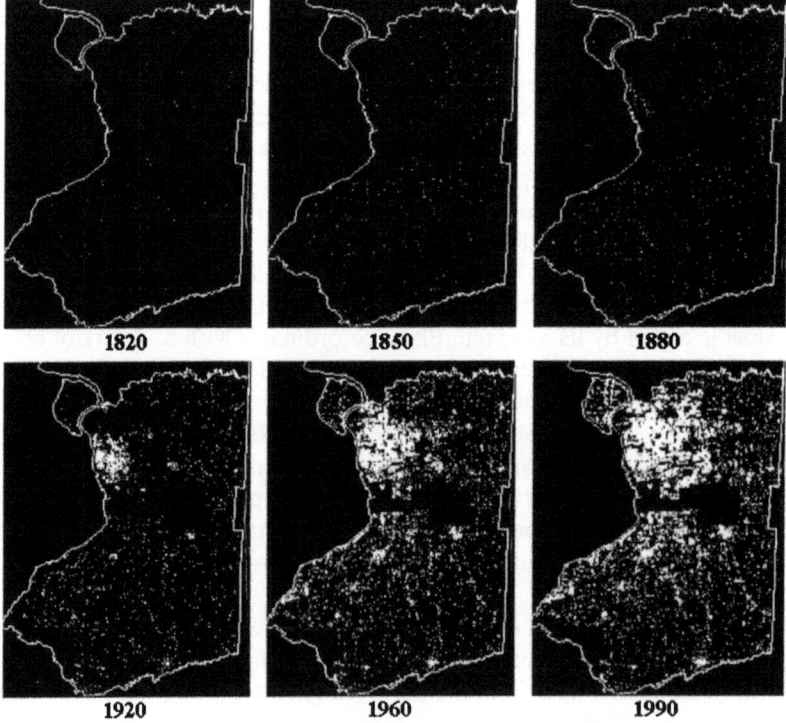

Figure 13.1 The growth of Buffalo, 1750–1990.

the conventional fashion. Until the late 1950s, the industrial base remained intact but from 1960, the core began to decline and manufacturing industry all but disappeared. The last 30 years have seen rapid urban decline in the core which is now spreading to the inner suburbs. The region is losing population and there is little sign that this will be stemmed. Continued exurbanisation and some effects from the Toronto region are factors which might dominate the future but the picture is unclear. An interpretation of these patterns in terms of urban theory is presented in Batty and Xie (1996) but these differ profoundly from the more superficial aggregate analyses which have dominated this field for the last 50 years. New data sources at the micro level enable much richer and informed analysis of patterns of urban development to which we now turn.

13.4 VISUALISING URBAN SPACE–TIME HISTORIES

Most statistical and visualisation methods have not been developed with very large data sets in mind, and thus there are major problems in using existing software. Statistical estimates are usually made for data containing hundreds, perhaps thousands of observations but not hundreds of thousands, and as one scales in this fashion, problems of noise, error and anomalous observations change in their significance to the analysis. Scale is of equal but different significance when it comes to the visualisation of large data sets. In terms of resolution, there is an immediate association between the density of data points in space and the spatial resolution of the display device on which the data is mapped. For standard workstations (based on the Sparc 2 for example), screen resolution is of the order of 800×1000 pixels, and thus data sets such as this one with 250 000 or so co-ordinate pairs can, in principle, be displayed with little loss of resolution. However, it is likely that the screen available for display will be perhaps only half or less than the total area available, and combined with the fact that the density of points will vary in most applications, there will be some considerable loss of resolution in problems of the size considered here. For example, in most of our displays of the Buffalo data, we rarely use more than 300×300 pixels and thus we are losing the resolution of the original data by at least four times. Lastly, the speed of display device if the screen is addressed directly from programs written in C, is slower than using special functions based on rasterising the data available in most GIS software.

To display, say 250 000 points directly in *ARC-INFO*, involves times that are unacceptable unless transformations and aggregations of the data are invoked. In the examples which follow, a pointgrid command is used to rasterise the data to different levels of spatial aggregation, and thus the GRID module enables the data to be displayed immediately but with problems which we will now explain. We have gridded the basic data points for the last year of the cumulative development (1989) to 5, 10, 20, 50, 100 and 200 m grids and we show these in Figure 13.2. We have also looked at 500 and 1000 m grids but these fuse the data too heavily for our purposes. Examining the 5 m grid display in Figure 13.2 reveals that there are only a couple of hundred pixels lit for a data set containing a quarter of a million points. In fact, when the 5 m data is displayed, a pixel is only lit if more than half the 5 m cells which comprise that pixel are occupied by the parcel data. In short, because the 5 m grid is very sparse at that level, most grid squares are empty. What we are seeing is a translation of this pattern of very low density of grid occupance directly to the

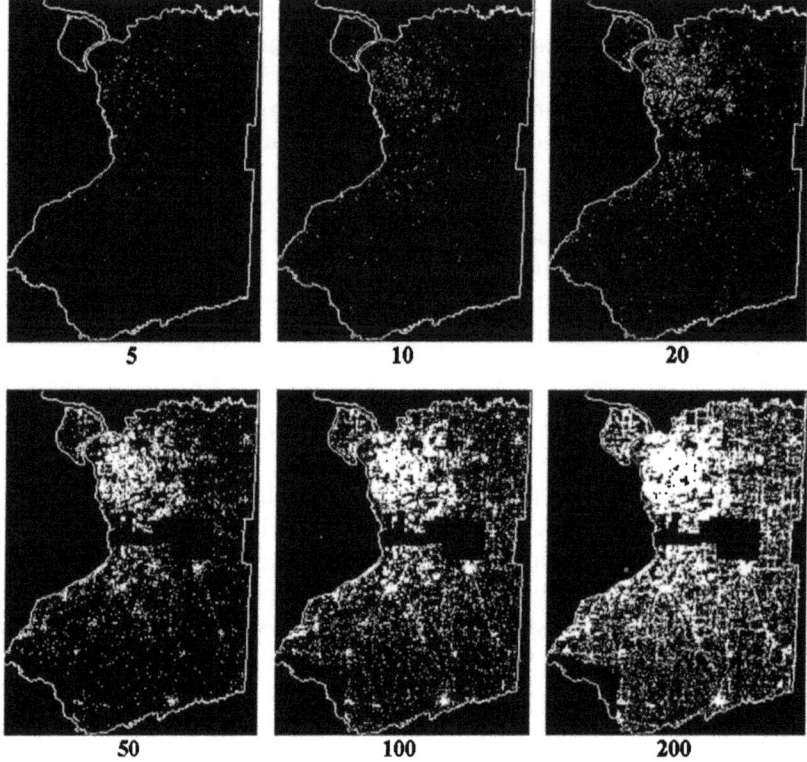

Figure 13.2 Variations in spatial resolution in metres.

screen. If we zoom in on this map within Figure 13.2 towards the 5 m scale, then the points reveal themselves illustrating the sparsity of the data. Such a zoom is shown for an area in the north-eastern suburbs in Figure 13.3 where raw TIGER data has been overlaid on top of the points to give a sense of the fact that we are viewing a residential area of the city. Figure 13.2 also shows how the density of points changes as we grid the data at successively higher levels of spatial resolution. From this, it would appear that the 100 m level provides the most visually acceptable level but this is in terms of 'our' perception of what we think urban-residential development actually is. Two hundred metres looks too dense and this whole process throws into stark relief the problem of what we mean by urban development, forcing us to directly consider the key issues in using micro data in the analysis of macro pattern. As each level of resolution provides a measure of the density, then one way forward would be to decide what types of density are being explored. For example, the usual definition of urban development would imply that the structure on the site and the lot itself would be 'urban' and thus this might imply that a grid of 100 m is an appropriate way of detecting this. If, however, one were looking at the density of structures themselves, then a grid of 20 or 30 m might be more appropriate. In this case, we have chosen the 100 m level to work with.

The level of temporal aggregation or resolution for this type of data is equally problematic. We have less control over the data in this dimension as the series we begin with is at 1800 ending at 1989 in yearly increments. We also need to concate-

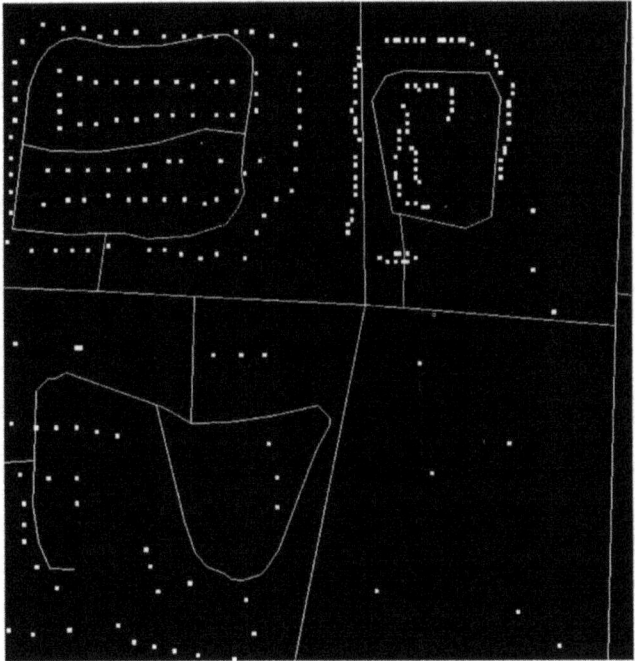

Figure 13.3 Zooming in to the 5 m level.

nate grid spacing with temporal aggregation and the same comments as those above apply with respect to density across time as well as space. Essentially we need to examine data at the lowest temporal level which is annual change, yet it is quite possible the patterns that we seek will not show up at this level of aggregation and thus we have explored the spatio-temporal resolution over 2, 5, 10, 20 and 50 year increments. For completeness, we have concatenated these 6 time resolutions with the 6 spatial resolutions but due to limits on space, we only show one slice through this matrix here at the 100 m level in Figure 13.4. We have taken the 6 time resolutions from the 1980 data which represents a period when the city was growing quite rapidly in the north-east and south-east. In Figure 13.4, the patterns of growth for the 1 year (1979–1980) and 2 year (1978–1980) increments are indistinct. The 5 year (1975–1980) pattern is clearer as is the 10 year (1970–1980), but the 20 year (1960–1980) and 50 year (1930–1980) seem less meaningful. If we examine the 20 m grid, a clear focus does not emerge until the 10 year increment while the 20 and 50 year aggregations also reveal more articulate structure than the same for the 100 m grid. What we have not done is to explore how each time increment might vary throughout the entire growth period. Moreover, it is entirely possible that patterns might emerge when fixed time increments are moved forwards or backwards but using the basic 1 year interval as the fixed anchor for each aggregation. In short, this could be seen as some sort of spatio-temporal moving-average being based on, say, exploring how the changing pattern over 5 year resolutions evolve; an example would be to construct the overlapping frames at (1975–1980), (1976–1981), (1977–1982) . . . and so on. It is essential to automate our exploration of how spatial patterns change. The sheer number of observations at the individual level makes possible analysis of continua and animation thus becomes possible. In the rest of

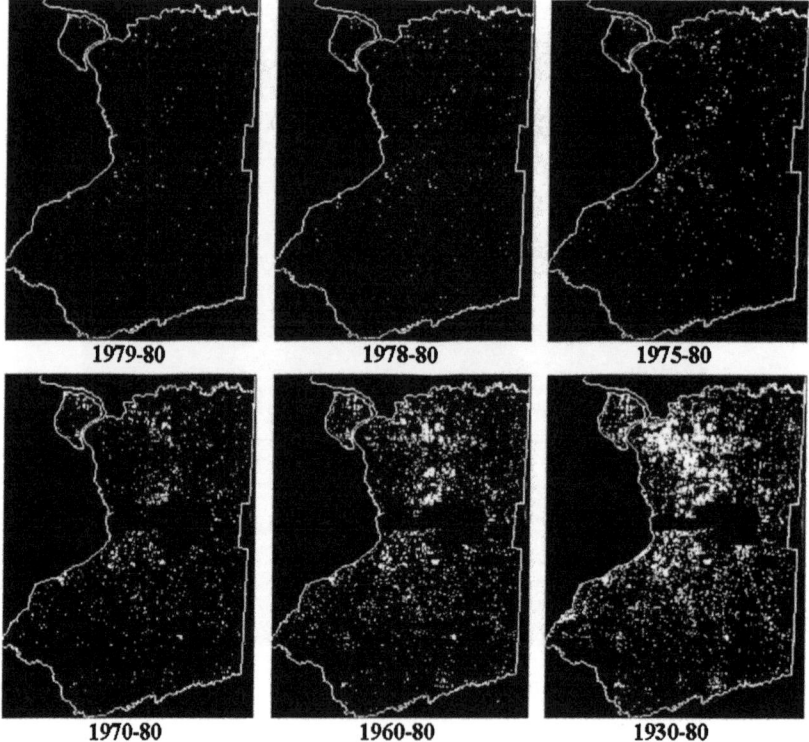

Figure 13.4   Variations in temporal resolution (100 m grid).

this chapter, we will explore how we might build a simple tool which will animate the data using time as the basic organising principle.

13.5   ANIMATING URBAN DYNAMICS

In constructing the animation, it is important to give some visual reference to the data so that errors can be examined when two or more data sets are combined. This is particularly important when parcel data is examined at the finer spatial scales rather than for the entire urban region. We have merged the parcel data set with two others which mainly involve physical features – topographic data from a digital elevation model (DEM), and line feature data pertaining to roads, rivers, and administrative boundaries. The first data set which we use as a backcloth for both positional as well as error analysis purposes is taken from the USGS 1 : 250 000 DEM model sheets available on the net. The second is the raw TIGER file data for the Buffalo region compiled by the Bureau of the Census and USGS, primarily for the display of the 1990 Census, and a version of this data from a third party data supplier (Claritas) from which we have extracted the Erie County boundary.

Our ability to relate these data sets is entirely due to the availability of GIS and related software which enables us to transform all the data to the same co-ordinate system (UTM-NAD23) and overlay the various coverages in a consistent manner. Our transformations may have introduced some error or at least proliferated error already in the positional data, although we have been surprised by how good the fits have been. We have used *ARC-INFO* for much of the work but the best overlay display capability for our purposes is within the image processing software *Imagine* from ERDAS while the Claritas Census data was originally in the GIS *MapInfo*.

The DEM data from USGS consists of four sheets which require being joined together, in this case using standard functions within *ARC-INFO*. The main problem we encountered was stitching the two 'Canadian' to the two 'US' sheets. It appears that up to five lines of pixels are missing here but in the vicinity of the US–Canadian border, the Niagara river is completely in error in the area of Grand Island which does not exist. There are procedures for correcting these types of error systematically but in this case, these errors were so severe that it was necessary to touch-up and import the 'correct' detail by hand, using Adobe *Photoshop*. In Figure 13.5, we have overlaid the parcel data on the touched-up DEM using *Imagine* which gives us better overlay capability for situations where some transparency of layers is necessary. Although the correspondence seems excellent, there are errors which emerge when we move to finer scales. When we zoom in on the area of Buffalo city itself, there are several locations/parcels which are located in Lake Erie. We consider these to be miscoded as the rectification error of the two images is not as great as the displacements of these parcels.

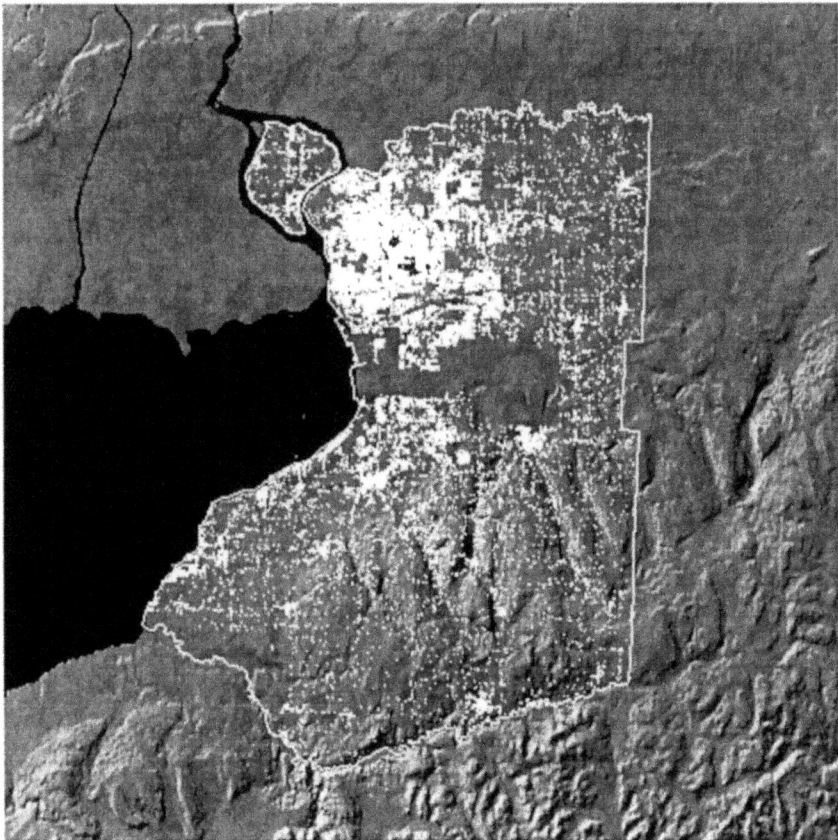

Figure 13.5   The property parcel data layered across the DEM.

We can construct our animations at any spatial scale and over whatever time increments are desired. In Figure 13.6, we have zoomed in on the North Campus area of Amherst in north-east Buffalo using the 1989 100 metre grid on which we have overlaid raw TIGER data. The campus area is clearly visible as the 'hole' in the centre of this map area bounded by the inner S-shaped peripheral road. The grid squares which are developed are shown in white. If we had access to color, we could have shown the age of development and in our many explorations of this data set, we have seen how easy it is to associate buildings of certain ages which we know to be correct, with the TIGER line file data, thus providing robust tests of the data. If we were to animate the data at this scale, then it would be important to use TIGER data to give the locational referent. Unfortunately, the TIGER data is only available for 1990, thus a more complete animation sequence involving the construction of new roads and so on along with changing urban development is not possible. Yet the possibilities for combining various aggregations across space and through time are endless and thus some structure needs to be imposed on the possible range of choices. It is possible to zoom in on any subdivision of the region at whatever scale and run the animation with the possibility of changing this zoom as the animation is in progress. Animations of urban growth have existed in various forms for decades. Indeed, the idea of overlay itself implies one idea of animation. Serious work in animating urban systems emerged, however, once computer graphics of

Figure 13.6 Zooming in on the North Campus.

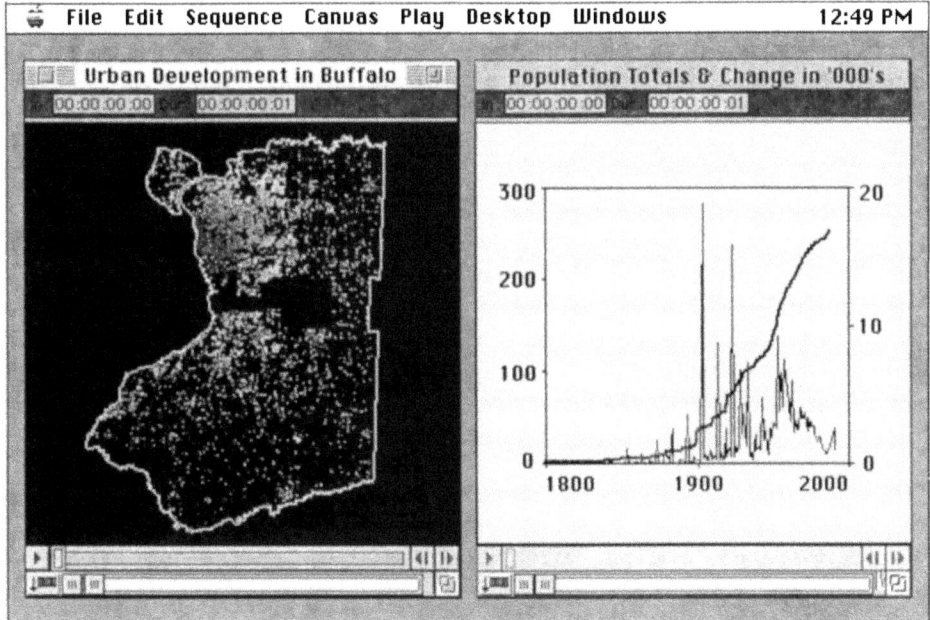

Figure 13.7 A mockup of the urban animator.

three-dimensional surfaces became popular, first illustrated in a prescient article by Tobler (1970). Animations based on real cities have, however, been somewhat stymied for lack of data although software for spatial–temporal interpolation and for scanning historical maps is opening up new opportunities. Since the 1960s, models of urban growth have used some form of animation in their display as static frames (for an example based on Buffalo, see Lathrop and Hamburg's (1965) work) but with the development of urban cellular automata and fractal models, computer simulations as animations are being widely applied (Engelen *et al.*, 1993; Batty and Longley, 1994).

A serious attempt based on 'real' time series data which we are unashamedly copying is the animation of the urban growth of the San Francisco Bay Area by a group at USGS (Kirtland *et al.*, 1994). From around 1850 until the 1940s, old maps form the source of the key frames, from the 1950s until the 1970s, land-use databases provide the frames, while since then, remotely sensed imagery has been used, the entire sequence being based on inbetweening these key frames. A movie has been constructed on the web and if you log-on to http://128.102.124.15/usgs/HILTStart/, then an mpeg movie can by launched by clicking on the first *hotlink* you come to. This clearly shows how the Bay Area has been surrounded by development and how the Sacramento region is now exploding in growth.

Our own efforts so far involve three approaches. Yichun Xie has developed an animation of the Buffalo data outside any proprietary GIS or RS software based on simple C code running in an X window and we are developing this application further. Our second approach has involved using the movie maker-player which exists within *Imagine*. So far, we have been able to simulate only 20 frames – the growth from 1969 to 1989 – due to constraints on the swap space imposed on our workstations although we should be able to animate the entire sequence of frames from 1800 to 1989 when our system is next reconfigured. This player enables the user to run forwards at different speeds, to freeze the frame and then move one frame at a time either forwards or backwards when required. Several movies can be loaded at once and from our preliminary runs we have already discovered that it is essential to look simultaneously at both cumulative and incremental urban development as the grid size does suppress growth if the cell is already occupied. We have not yet managed to run two movies in phase but we consider that this is possible.

However, the best environment we have found for animation is using software on the Macintosh platform. We have imported the time series into *MacroMind Director*, and have developed various simulations to provide some sense of the forward evolution and backward devolution of the region. What we are now doing is computing several spatial statistics – averages, dimensions, totals and so on from the spatial patterns showing how these change using graphs alongside the simulation. In Figure 13.7 we show a mockup of the kind of interface we are developing using *Videoshop* running on a Mac.

13.6 NEXT STEPS

Modelling urban development involves beginning with theory and then testing that theory against some data in the quest to validate and improve the original hypotheses. We began with brief speculations based on some simple but profound hypotheses, and then merely implied that our data was consistent with those hypotheses. However, to demonstrate this we began a detailed exploration of the data and the ways it might be used to study urban growth. We have thus ended up a long way from our original hypotheses but we believe that the base we are building is essential if these hypotheses concerning urban change are to be tested in any complete and serious way.

The theory of urban development based on self-organised criticality can be tested in many different ways. The most obvious and superficial establishing measures of consistency between theory and data, have been noted earlier and these involve showing that the urban system has fractal properties which change but then remain constant through time. These measures can be developed in countless ways. We also intend to develop other ways of animating our data, building simple models of interpolation based on the time series and following the Bay Area example of morphing change from other data sources such as RS imagery and old maps (Batty and Howes, 1995). Building a more complete and comprehensive time series by fusing our parcel data with these independent data sources is another goal, while developing different kinds of animations in other software will lead to new insights into the data and the problem. If you wish to read more extended reports on this research, log-on to http://www.geog.ucl.ac.uk/casa/urs.html *or* http://www.geog. buffalo.edu/ and follow the menu to *The Buffalo Project*.

REFERENCES

ALLEN, P. M. (1994) *Cities and Regions as Self-organizing Systems: Models of Complexity*, International Ecotechnology Research Centre. Bedford, UK: Cranfield University.
BAK, P. and CHEN, K. (1991) Self-organized criticality. *Scientific American* **264**, 4653.

BATTY, M. and HOWES, D. (1995) Predicting temporal pattern in urban development from remote imagery. A paper presented at the European Science Foundation GISDATA Seminar on Remote Sensing and Urban Change, Strasbourg, France, June (see the web page http://www.shef.uk.ac/uni/academic/DH/gis/remsens.html).

BATTY, M. and LONGLEY, P. (1994) *Fractal Cities: a Geometry of Form and Function.* London and San Diego, CA: Academic Press.

BATTY, M. and XIE, Y. (1994) From cells to cities. *Environ. Plan.* B, **21**, s31–s48.

BATTY, M. and XIE, Y. (1996) Self-organized criticality and urban development. (Submitted to *Regional Studies*).

DENDRINOS, D. (1992) *The Dynamics of Cities: Ecological Determinism, Dualism and Chaos.* London and New York: Routledge.

ENGELEN, G., WHITE, R. and ULJEE, I. (1993) Exploratory modelling of the socioeconomic impacts of climate change. In MAUL, G. A. ed., *Climatic Change in the Intra-Americas Sea.* London: Edward Arnold, pp. 351–368.

KIRTLAND, D., GAYDOS, L., CLARKE, K., DE COLA, L., ACEVEDO, W. and BELL, C. (1994) An analysis of transformations in the San Francisco/Sacramento area. *World Resource Rev.* **6**, 206–217.

KRUGMAN, P. (1994) Complex landscapes in economic geography. *Am. Econ. Assoc. Pap. Proc.* **84**, 412–416.

LATHROP, G. T. and HAMBURG, J. R. (1965) An opportunity-accessibility model for allocating regional growth. *J. Am. Inst. Plan.* **31**, 95–103.

TOBLER, W. (1970) A computer movie simulating urban growth in the Detroit region. *Econ. Geogr.* **26**, 234–240.

WHITE, R. and ENGELEN, G. (1993) Cellular automata and fractal urban form: a cellular modelling approach to the evolution of urban land use patterns. *Environ. Plan.* A, **25**, 1175–1193.

WILSON, A. G. (1981) *Catastrophe Theory and Bifurcation: Applications to Urban and Regional Analysis.* Berkeley, CA: University of California Press.

CHAPTER FOURTEEN

Scale-based characterisation of digital elevation models

JOSEPH WOOD

14.1 INTRODUCTION

Digital elevation models (DEMs) are often used within a GIS environment to characterise elements of the real surface they represent. Characterisations may be of relatively low-level properties, here referred to as morphometric parameters, such as slope magnitude, slope direction, surface convexity (e.g. Evans, 1972, 1980). These parameters may be used to identify morphometric features such as local pits, peaks, ridges, channels, and passes (e.g. Peucker and Douglas, 1974). Alternatively, morphometric features and parameters may be combined or used directly to characterise higher level geomorphological properties such as drainage basins, drainage networks, and valley heads (e.g. O'Callaghan and Mark, 1984; Band, 1989; Tribe, 1990).

The large literature on DEM characterisation has tended to concentrate on the computational process of extracting morphometric parameters and features from gridded elevation models. A number of papers have evaluated the success of such procedures both empirically and theoretically. For example, Skidmore (1989) compared the slope values calculated from a DEM using a variety of common methods with the 'true' slope value found by manually examining contour information. Douglas (1986) examined the effectiveness of the procedures for identifying ridges and channels. Carter (1992) looked at the effect of data precision on slope and aspect identification.

While there have been undoubted improvements in automated DEM characterisation over the past two decades, there are a number of, as yet, unresolved problems. These are based upon the failure of most procedures to take into account the effect of scale on the morphometric properties of a surface. Most existing methods for calculating morphometric parameters are based upon passing a local (usually 3 by 3) kernel across a DEM (Skidmore, 1989). However, what is not always acknowledged is that information derived from such kernels is only relevant to the *scale implied by the resolution of the DEM*. This scale may not even be directly related to the grid cell resolution (Hodgson, 1995). Since this scale is often arbitrary and not necessarily related to the scale of characterisation required, derived results may not always be appropriate.

This chapter considers methods of morphometric characterisation that involve a multi-scale description of surface form. After presenting a justification for this approach, a method of multi-scale characterisation is considered. Multi-scale quadratic approximation is shown to have use in generalisation of elevation models, and in the derivation of meaningful morphometric parameters and features.

14.2 JUSTIFICATION OF SCALE-BASED CHARACTERISATION

Geomorphology is multi-scale

Geomorphological characterisation of landscape involves the appreciation of surface form at a number of scales. Both our own geomorphological interpretation as well as the features themselves are defined across multiple scales. Despite the notion of scale independence (or 'self-similarity') suggested by the use of fractal geometry (e.g. Burrough, 1981), most geomorphologists would acknowledge for example the form and processes of rill development at the exterior of a drainage network are very different to the form and processes operating at the estuarine outlet of the same network. Several geomorphological studies have explicitly considered the role of scale in the characterisation of surface form (e.g. Church and Mark, 1980), but few have applied this consideration to DEM analysis.

Uncertainty in elevation models

Uncertainty can be introduced into DEMs for a number of reasons, but the form of that uncertainty is usually characterised by local, high frequency, variation (Wood and Fisher, 1993). Any process that attempts to characterise a surface only at the (high frequency) resolution of the DEM will be most susceptible to that uncertainty. Low frequency variation is not sufficient to fully describe the characteristics of a surface as only trends will be indicated. Therefore it is necessary to combine the scales of characterisation in order to distinguish real from erroneous surface pattern.

Addition of meaning to elevation models

Scale dependencies are implicitly coded within DEMs. By making such relations explicit, new information can be revealed. This may be to help classify surface form, or to explain the processes that give rise to such form.

Interpolation and generalisation

GIS allow data from a variety of scales to be integrated as well as allowing changes between scales. In such an environment, intelligent scale-based generalisation and interpolation can reduce errors associated with such transformations. Scale-based characterisation forms the first part of both generalisation and interpolation.

14.3 EXISTING TECHNIQUES FOR SCALE-BASED CHARACTERISATION

Although the techniques described in this work are new, they should be placed in the context of existing scale-based characterisation methods. Transformation to the frequency domain using *Fourier transforms* allows pattern to be modelled as a series of power spectra (Gonzalez and Woods, 1992). The transformation involves identifying the frequency of variation throughout the original surface and thus provides an explicit map of variation with respect to scale. This is one of the few techniques to explicitly consider scale variation, but the frequency transformation makes geomorphological interpretation difficult.

The fundamental property of *spatial autocorrelation* can be explicitly related to scale in the form of the variogram or corellogram. The nature of this relationship provides a (one-dimensional) description of scale-based variation (Goodchild, 1986). This representation can be used as the basis of interpolation (most notably in Kriging), but has had relatively little application in geomorphometrical interpretation. Alternatively, *fractal geometry*, although based on the assumption of scale invariance, can be considered over a limited range of scales and a picture of scale dependence can be built up (e.g. Culling, 1989). This is most succinctly represented in the form of so-called 'Richardson plots' (Goodchild and Mark, 1987).

A number of texture analysis techniques used in image processing may be adapted to look at 'textures' at different scales. Of particular use is the *co-occurrence matrix* and the measures that may be taken to describe them (Haralick *et al.*, 1973). These matrices may be visualised directly to give an impression of scale dependent texture (Wood and Fisher, 1993; Dykes, 1994). This process can be regarded as a specific instance of more general *adaptive kernel techniques*. Existing local kernel based operations (e.g. Evans, 1980) are applied over a number of scales determined by re-sampling a raster grid or extending the kernel size. Alternatively, some form of adaptive filter that changes size according to local DEM variation may be used (e.g. Skidmore, 1990).

14.4 MULTI-SCALE QUADRATIC APPROXIMATION

14.4.1 Local quadratic approximation

Several authors have shown that in order to calculate the morphometric parameters *slope, aspect, profile convexity* and *plan convexity* there is no more effective method than quadratic approximation using least squares fitting (e.g. Skidmore, 1989). Here, a local 3×3 window or kernel is passed over the DEM from which six quadratic coefficients may be calculated that describe a second order trend surface in the form:

$$z = ax^2 + by^2 + cxy + dx + ey + f \tag{14.1}$$

The six coefficients are found from the nine (overspecified) sample points using least squares. That is, a quadratic expression is found with minimal squared deviation from the nine sample points. Evans (1979) shows that due to the regular nature of the window in the raster co-ordinate system, the coefficients can be expressed for the 3×3 case, as follows. For a 3×3 window of grid resolution g, elevation values z_1 to z_9, z_5 being the central cell:

z_1	z_2	z_3
z_4	z_5	z_6
z_7	z_8	z_9

$$a = (z_1 + z_3 + z_4 + z_6 + z_7 + z_9)/6g^2 - (z_2 + z_5 + z_8)/3g^2$$

$$b = (z_1 + z_2 + z_3 + z_7 + z_8 + z_9)/6g^2 - (z_4 + z_5 + z_8)/3g^2$$

$$c = (z_3 + z_7 - z_1 - z_9)/4g^2$$

$$d = (z_3 + z_6 + z_9 - z_1 - z_4 - z_7)/6g$$

$$e = (z_1 + z_2 + z_3 - z_7 - z_8 - z_9)/6g$$

$$f = (2(z_2 + z_4 + z_6 + z_8) - (z_1 + z_3 + z_7 + z_9) + 5z_5)/9$$

(from Evans (1979) p. 29)   (14.2)

Although this method can be regarded as a maximum likelihood procedure, it is only so at the scale defined by the resolution of the kernel. That is, it only describes morphometric variation at a wavelength of approximately $3g$. Clearly, most surfaces will contain variation at other scales, and so what follows is a generalisation of the least squares procedure for any size of 'local' window.

14.4.2   Generalised quadratic approximation

To solve an overspecified quadratic expression using least squares, the six coefficients of the quadratic (equation 14.1) are expressed as six simultaneous equations, or *normal equations*:

$$a \sum x_i^4 + b \sum x_i^2 y_i^2 + c \sum x_i^3 y_i + d \sum x_i^3 + e \sum x_i^2 y_i + f \sum x_i^2 = \sum z_i \cdot x_i^2$$

$$a \sum x_i^2 y_i^2 + b \sum y_i^4 + c \sum x_i y_i^3 + d \sum x_i y_i^2 + e \sum y_i^3 + f \sum y_i^2 = \sum z_i \cdot y_i^2$$

$$a \sum x_i^3 y_i + b \sum x_i y_i^3 + c \sum x_i^2 y_i^2 + d \sum x_i^2 y_i + e \sum x_i y_i^2 + f \sum x_i y_i = \sum z_i \cdot x_i y_i$$

$$a \sum x_i^3 + b \sum x_i y_i^2 + c \sum x^2 y_i + d \sum x_i^2 + e \sum x_i y_i + f \sum x_i = \sum z_i \cdot x_i$$

$$a \sum x_i^2 y_i + b \sum y_i^3 + c \sum x_i y_i^2 + d \sum x_i y_i + e \sum y_i^2 + f \sum y_i = \sum z_i \cdot y_i$$

$$a \sum x_i^2 + b \sum y_i^2 + c \sum x_i y_i + d \sum x_i + e \sum y_i + fN = \sum z_i$$

(14.3)

where a to f are the coefficients as in equation 14.1, x_i and y_i are the planimetric co-ordinates of the centre of each cell in the window, z_i is the elevation of each cell in the window, and N is the number of window cells.

This expression can be simplified by adopting a local co-ordinate system with an origin at the centre of the window, and grid cell resolution of g. Thus, for a 5×5 window, the co-ordinate system would become:

$(-2g, -2g)$	$(-2g, -g)$	$(-2g, 0)$	$(-2g, g)$	$(-2g, 2g)$
$(-g, -2g)$	$(-g, -g)$	$(-g, 0)$	$(-g, g)$	$(-g, 2g)$
$(0, -2g)$	$(0, -g)$	$(0, 0)$	$(0, g)$	$(0, 2g)$
$(g, -2g)$	$(g, -g)$	$(g, 0)$	(g, g)	$(g, 2g)$
$(2g, -2g)$	$(2g, -g)$	$(2g, 0)$	$(2g, g)$	$(2g, 2g)$

The symmetry of the local co-ordinate system reduces all expressions in equation 14.3 without even exponents throughout, to zero. In addition, if the assumption of square grid cell size is adopted, the sum of x^2 is equivalent to the sum of y^2, and the sum of x^4 is equivalent to the sum of y. For computational reasons, it is convenient to express the normal equations in matrix form (Unwin, 1975). Thus, the simplified normal equations are represented as:

$$\begin{bmatrix} \sum x_i^4 & \sum x_i^2 y_i^2 & 0 & 0 & 0 & \sum x_i^2 \\ \sum x_i^2 y_i^2 & \sum x_i^4 & 0 & 0 & 0 & \sum x_i^2 \\ 0 & 0 & \sum x_i^2 y_i^2 & 0 & 0 & 0 \\ 0 & 0 & 0 & \sum x_i^2 & 0 & 0 \\ 0 & 0 & 0 & 0 & \sum x_i^2 & 0 \\ \sum x_i^2 & \sum x_i^2 & 0 & 0 & 0 & N \end{bmatrix} \cdot \begin{bmatrix} a \\ b \\ c \\ d \\ e \\ f \end{bmatrix} = \begin{bmatrix} \sum z_i \cdot x_i^2 \\ \sum z_i \cdot y_i^2 \\ \sum z_i \cdot x_i y_i \\ \sum z_i \cdot x_i \\ \sum z_i \cdot y_i \\ \sum z_i \end{bmatrix} \qquad (14.4)$$

The four unique elements of the matrix of sums of squares and cross products may be found efficiently for any window size of n^2 cells with a single pass involving approximately $2n$ calculations. The matrix expression can be solved using LU decomposition and LU back substitution (Press et al., 1988) to give the six quadratic coefficients for any window size, and thus a quadratic characterisation over a range of scales.

14.4.3 Quadratic generalisation

It has been recognised that the overspecification of the quadratic function using nine sample points results in a degree of terrain generalisation (e.g. Evans, 1980; Zevenbergen and Thorne, 1987). While this may not always be desirable for terrain characterisation, it does provide a mechanism for surface generalisation, and in particular, terrain smoothing.

Smoothing of elevation models to eliminate unwanted or spurious detail is frequently accomplished by using some form of mean filtering (e.g. O'Callaghan and Mark, 1984; Weibel and Heller, 1991). This has the desired effect of reducing the amplitude of high frequency elevation change, but it also has the undesired consequence of reducing the variance of the entire DEM.

An alternative method is proposed here using least squares fitting described above. The coefficient f in equation 14.1, above, gives the modelled elevation at the origin of the quadratic without necessarily drawing values towards the mean. For larger window sizes, the degree of overspecification of the quadratic is increased, thus increasing generalisation of the surface while maintaining the optimal quadratic characterisation. This provides the basis for effective terrain smoothing.

To investigate the effectiveness of quadratic generalisation as a device for terrain smoothing, a comparison was made with the more common method of mean filtering. Three contrasting DEMs were selected for processing. Lakes.dem is a 700×700 DEM of the English Lake District selected from tiled Ordnance Survey 50 m resolution data. The high relative relief is reflected in a mean elevation of 316 m and standard deviation of 188 m. Leics.dem is a 270×270, 50 m DEM of a part of Leicestershire again selected from Ordnance Survey 50 m resolution DEM data.

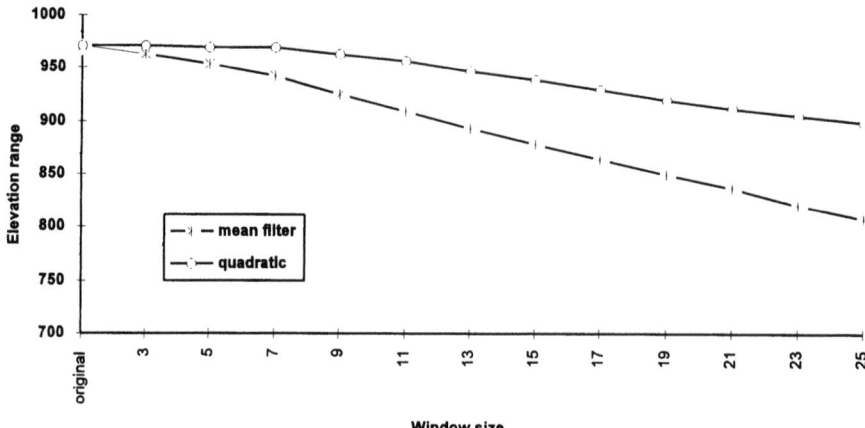

Figure 14.1   Effect of smoothing on the elevation range of Lakes.dem.

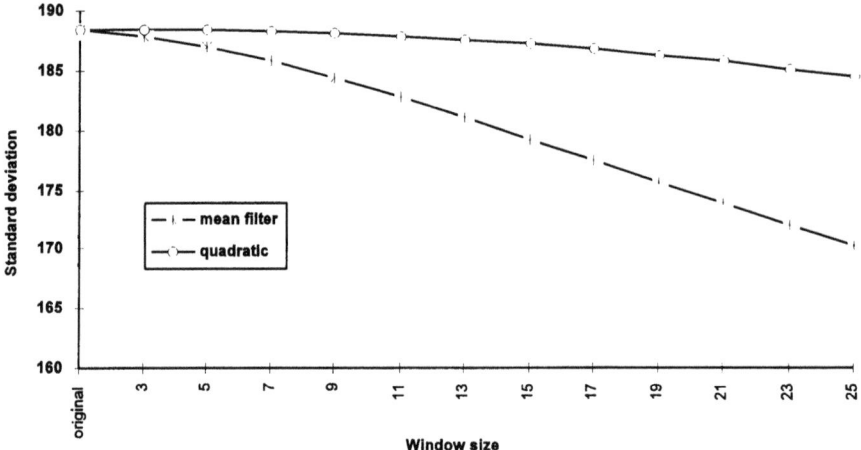

Figure 14.2   Effect of smoothing on the standard deviation of Lakes.dem.

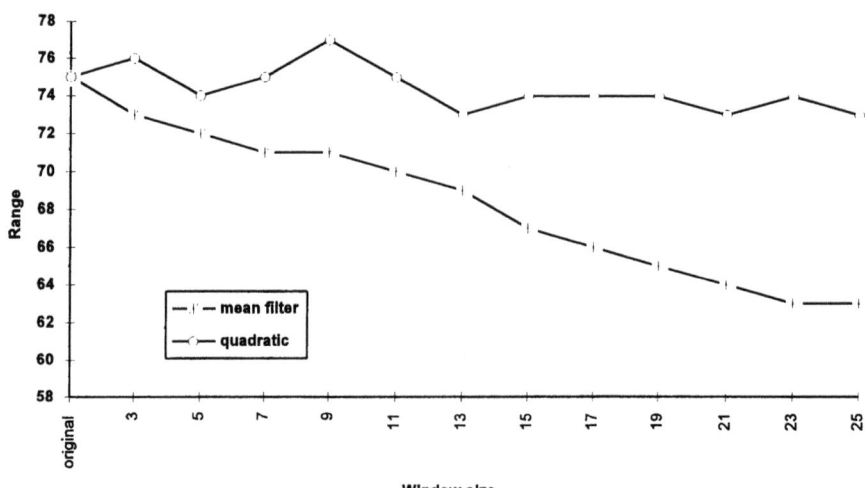

Figure 14.3   Effect of smoothing on the elevation range of Leics.dem.

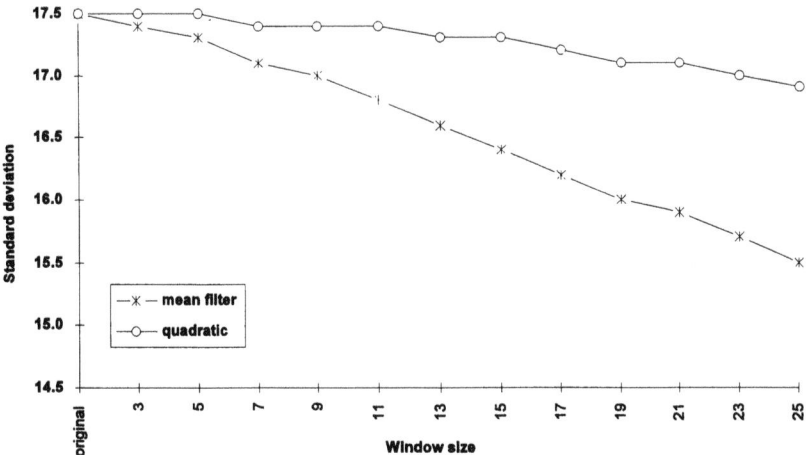

Figure 14.4   Effect of smoothing on the standard deviation of Leics.dem.

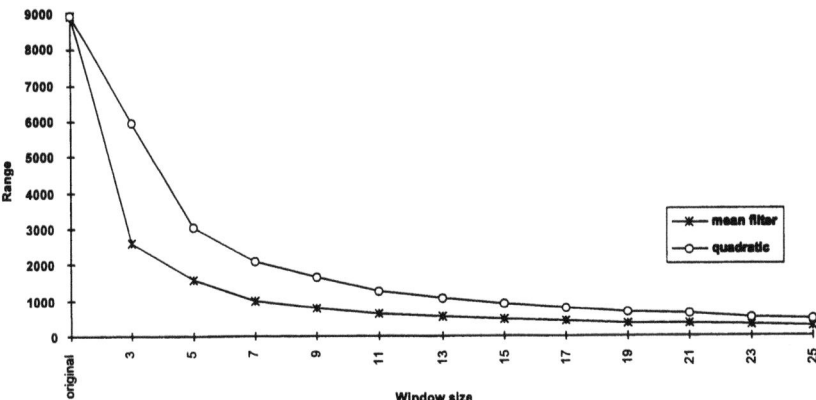

Figure 14.5   Effect of smoothing on the elevation range of Gauss.dem.

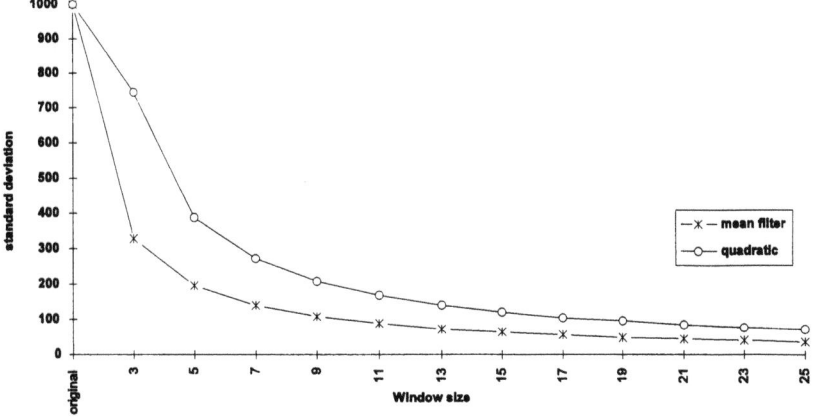

Figure 14.6   Effect of smoothing on the standard deviation of Gauss.dem.

Figure 14.7   Lakes.dem unsmoothed.

The much lower relative relief is reflected in a mean elevation of 46 m and standard deviation of 18 m. Finally, a non-spatially correlated gaussian surface of 200 by 200 cells, Gauss.dem, was created with mean value of 0 and standard deviation of 1000.

Figure 14.8   Lakes.dem after 25 × 25 (1.25 km) quadratic.

Both mean filtering and quadratic approximation techniques were applied to all three surfaces using window sizes from 3×3 to 25×25 cells. Mean filtering was achieved by calculating the unweighted mean of all cells within the window and storing the result at the co-ordinates of the centre. The mean, standard deviation and range of each DEM were recorded after smoothing at each scale. The results of the range and standard deviation values are shown in Figures 14.1–14.6.

Both forms of smoothing have an effect on the standard deviation and range of elevation values within the DEM, but little effect on mean elevation (which remained within 0.1 m of the original at all stages). However, mean filtering degrades both dispersion measures to a far greater extent than quadratic approximation. As would be expected, degradation is greatest for the uncorrelated surface and least for the highly spatially autocorrelated surface of Leicestershire. The uncorrelated surface shows a convergence of dispersion measures as the window size is increased, with a maximum difference at small window sizes. For the two 'real' surfaces, the difference between quadratic approximation and mean filtering appears to increase with degree of smoothing (window size). This would suggest that quadratic approximation has particular advantages for the production of highly smoothed surfaces.

Figures 14.7 and 14.8 show the effect of applying this method to the Lakes.dem using a 25×25 (1.25 km) filter.

14.5 MULTI-SCALE QUADRATIC FEATURE DETECTION

The more widely employed application of local window characterisation is in the derivation of morphometric parameters and morphometric features. Parameters of slope and aspect may be calculated directly from the six quadratic coefficients as reported by Evans (1980) and Skidmore (1989):

$$\text{slope} = \arctan(\sqrt{d^2 + e^2})$$

$$\text{aspect} = \arctan\left(\frac{e}{d}\right) \tag{14.5}$$

By varying the size of window used to derive the quadratic coefficients, slope and aspect may be measured at a variety of scales.

14.5.1 Measurement of convexity

The measurement of the second derivative of elevation – surface curvature is useful in that it can be strongly related to geomorphological process. However, its derivation is somewhat more problematic than slope or aspect since the partial derivative of the bi-variate quadratic yields an infinite number of solutions. Surface convexity must therefore be expressed in some defined direction. Evans (1979) separates curvature into two orthogonal components – *profile* and *plan convexity*, the former being the vertical component in the direction of aspect, the latter being the horizontal component. These measures can only be calculated if the slope normal is not vertical. For this special case, two alternative measures are used – minimum and maximum curvature (Evans, 1979).

```
Name:          Mfeature(coeff,n)
Purpose:       Identifies the morphometric feature described by a bivariate quadratic
               function.
Parameters:    coeff[6]      Array holding the 6 quadratic coefficients a to f
               n             Size of side of local window
Globals:       Tslope Minimum gradient that defines a non flat surface
               Tconvex       Minimum convexity defining a non-planar surface
               g             Grid resolution of DEM
Locals :       slope         Maximum gradient
               crosc         Cross-sectional convexity
               minic,maxic   Minimum and maximum convexity
```

```
Mfeature(coeff)
{
        /* Measure morphometric parameters */
        slope = atan(sqrt(d*d + e*e))
        crosc = n*g*(b*d*d + a*e*e - c*d*e)/(d*d + e*e)
        maxic = n*g*(-a-b+sqrt((a-b)*(a-b) + c*c))
        minic = n*g*(-a-b-sqrt((a-b)*(a-b) + c*c))

        /* Identify morphometric features */
        if (slope > Tslope)            /* Case 1: Surface is sloping */
        {
            if (crosc > Tconvex)
                return (RIDGE)
            if (crosc < -Tconvex)
                return (CHANNEL)
            else
                return (PLANAR)
        }
                                       /* Case 2: Surface is horizontal */
        if (maxic > Tconvex)
        {
            if (minic > Tconvex)
                return (PEAK)
            if (minic < -Tconvex)
                return (PASS)
            else
                return (RIDGE)
        }

        if (minic < -Tconvex)
        {
            if (maxic < -Tconvex)
                return (PIT)
            else
                return (CHANNEL)
        }
}
```

Figure 14.9   Algorithm for detecting morphometric features.

For the purposes of this study, an alternative measure of convexity is used –
cross-sectional convexity crossc. This is the curvature in the downslope perpendicular direction of the slope vector. As a measure it has the advantage of being computationally simpler to calculate than Evans' profile or plan convexity, and can be
directly related to geomorphological form when surface concavities are detected
(channel cross-section). Like profile convexity, this measure cannot be calculated
when the slope normal is vertical; in this special case minimum and maximum curvature are calculated (Evans (1979) p. 33).

$$\text{crosc} = -2\left(\frac{bd^2 + ae^2 - cde}{d^2 + e^2}\right)$$

$$\text{minic} = -a - b - \sqrt{(a-b)^2 + c^2}$$

$$\text{maxic} = -a - b + \sqrt{(a-b)^2 + c^2}$$

(14.6)

Surface curvature values were calculated for a 200×200 subset of Lakes·dem at the 12 window scales ranging from 3×3 cells (150 m) to 25×25 cells (1.25 km). The results of these calculations were visualised using a bi-polar colour scheme with the hues blue and red indicating concavity and convexity respectively and intensity indicating the strength of curvature. The twelve images were then viewed as an animated sequence to give a single 'visualisation' of the scale dependency of surface curvature. Selected views of this animated sequence are shown in Plate 2 (see colour plate section).

Several important scale dependent characteristics are made explicit by this visualisation. First, ridges tend to be expressed at a finer scale to valleys (typically at $w = 9$ for ridges, $w = 21$ for valleys). This is typical of an upland glaciated landscape such as the Lake District, and may be contrasted with previously periglacial upland environments such as Dartmoor. Second, the network of ridges and valleys at the finest scale is far more fragmented than its expression at coarse scales. This can be interpreted as an increase in the statistical 'signal to noise' ratio with a change in scale.

14.5.2 Morphometric features

The morphometric parameters of slope and curvature may be combined to identify the morphometric features that comprise any surface. All parts of a surface may be classified into either a *peak, ridge, pass, channel, pit* or *plane*. This classification is important not only in its geomorphological significance, but as the basis of topological modelling of surface (Wolf, 1991) and the conversion to a TIN model (Fowler and Little, 1979). The algorithm used here for feature identification is a modification of that proposed by Evans (1979), and is shown in Figure 14.9.

A similar animated sequence was created from the 12 images of morphometric feature classification to give a single visualisation of scale dependency. Plate 3 (see colour plate section) shows a 'frame' from this sequence with all identified peaks named. The fact that these peaks do have names, most which are familiar to anyone who has walked in the region, demonstrates that this procedure identifies peaks that have morphometric significance at a scale that could not be identified at the 50 m resolution of the DEM. Thus, *Great Gable* can be identified as a single morphometric object at one scale, but also as a series of smaller peaks (e.g. *Green Gable, The Napes* etc.) at another.

14.6 CONCLUSIONS AND FURTHER RESEARCH

Geomorphologists have always considered the shape of land at a variety of scales. Our appreciation of what makes up the morphology of the landscape is inherently multi-scale. This work has attempted to show how we can encode some of that scale dependent information when automatically processing elevation models. Traditional methods of DEM processing have been extended to allow analysis of scale dependency, based primarily around the use of least squares fitting of a bivariate quadratic function at a variety of scales.

Characterisation of an elevation model is important in a variety of GIS contexts. Quadratic approximation has been shown to be an improvement over mean filtering as a mechanism for terrain model smoothing. Smoothing itself is useful for the basis of re-sampling at different scales, for the calculation of more realistic shaded relief, and as the basis of terrain visualisation. Characterisation of landscape properties through the measurement of morphometric parameters is a necessary part of many DEM based applications ranging from hydrological characterisation to landslide risk assessment. A treatment of the terrain at multiple scales can only be beneficial to such studies.

The ideas and techniques presented here can be most usefully extended in the field of morphometric feature identification. The identification of peaks at scales other than that defined by the resolution of the DEM has shown that it is possible to automatically identify features in the landscape that more closely relate to our perception of significance. It provides a mechanism for automated name placement of physically expressed features. Feature identification is the first step required to convert to alternative surface models. Selection of vertices for both the TIN model and topological weighted graphs can both be facilitated by such scale based characterisation.

REFERENCES

BAND, L. E. (1989) A terrain based watershed information system. *Hydrolog. Process.* **3**, 151–162.

BURROUGH, P. A. (1981) Fractal dimensions of landscapes and other environmental data. *Nature* **294**, 240–242.

CARTER, J. (1992) The effect of data precision on the calculation of slope and aspect using gridded DEMs. *Cartographica* **29**, 22–34.

CHURCH, M. and MARK, D. (1980) On size and scale in geomorphology. *Prog. Phys. Geogr.* **4**, 342–390.

CULLING, W. E. H. (1989) The characterization of regular/irregular surfaces in the soil-covered landscape by gaussian random fields. *Computers Geosci.* **15**, 219–226.

DOUGLAS, D. H. (1986). Experiments to locate ridges and channels to create a new type of digital elevation model. *Cartographica* **23**, 29–61.

DYKES, J. A. (1994) Area-value data: new visual emphases and representations. In HEARNSHAW and UNWIN eds, *Visualization in Geographical Information Systems*. Chichester: Wiley, pp. 103–114.

EVANS, I. S. (1972) General geomorphometry, derivatives of altitude, and descriptive statistics. In CHORLEY, R. J. ed., *Spatial Analysis in Geomorphology*. London: Methuen, pp. 17–90.

EVANS, I. S. (1979) An integrated system of terrain analysis and slope mapping. Final report on grant DA-ERO-591-73-G0040, University of Durham, England. p. 192.

EVANS, I. S. (1980) An integrated system of terrain analysis and slope mapping. *Z. Geomorphol.* Suppl-Bd 36, 274–295.

FOWLER, R. J. and LITTLE, J. J. (1979) Automatic extraction of irregular network digital terrain models. *Proceedings of ACM Siggraph '79, Chicago, IL, Computer Graphics* **13**(2), 199–207.

GONZALEZ, R. C. and WOODS, R. C. (1992) *Digital Image Processing*. Reading, MA: Addison-Wesley.

GOODCHILD, M. (1986) *Spatial Autocorrelation, Concepts and Techniques in Modern Geography (CATMOG) 47*. Norwich: GeoBooks.

GOODCHILD, M. F. and MARK, D. M. (1987) The fractal nature of geographic phenomena. *Ann. Assoc. Am. Geogr.* **77**, 265–278.

HARALICK, R. M., SHANMUGAM, K. and DINSTEIN, I. (1973) Textural features for image classification. *IEEE Trans. Syst. Man Cybern.* **3**, 610–611.

HODGSON, M. E. (1995) What cell size does the computed slope/aspect angle represent? *Photogram. Eng. Remote Sensing,* **61**, 513–517.

O'CALLAGHAN, J. F. and MARK, D. M. (1984) The extraction of drainage networks from digital elevation data. *Computer Vision, Graph. Image Process.* **28**, 323–344.

PEUCKER, T. K. and DOUGLAS, D. H. (1974) Detection of surface specific points by local parallel processing of discrete terrain elevation data. *Computer Graph. Image Process.* **4**, 375–387.

PRESS, W. H., FLANNERY, B. P., TEUKOLSKY, S. A. and VETTERLING, W. T. (1988) Numerical recipes. In *C – The Art of Scientific Computing.* Cambridge: Cambridge University Press, pp. 94–98.

SKIDMORE, A. K. (1989) A comparison of techniques for calculating gradient and aspect from a gridded digital elevation model. *Internat. J. Geogr. Inform. Syst.* **3**, 323–334.

SKIDMORE, A. K. (1990) Terrain position as mapped from a gridded digital elevation model. *Internat. J. Geogr. Inform. Syst.* **4**, 33–49.

TRIBE, A. (1990) Towards the automated recognition of landforms (valley heads) from digital elevation models. Proceedings of the 4th International Symposium on Spatial Data Handling, 1990, Zürich, Switzerland. 1, pp. 45–52.

UNWIN, D. (1975) An introduction to trend surface analysis. *Concepts and Techniques in Modern Geography (CATMOG)* **5**, Norwich: GeoBooks.

WEIBEL, R. and HELLER, M. (1991) Digital terrain modelling. In MAGUIRE, D. J., GOODCHILD, M. F. and RHIND, D. W. eds, *Geographical Information Systems: Principles and Applications.* London: Longman, pp. 269–297.

WOLF, G. W. (1991) A Fortran subroutine for cartographic generalisation. *Computers and Geosciences,* **17**(10), 1359–1381.

WOOD, J. and FISHER, P. (1993) Assessing interpolation accuracy in elevation models. *IEEE Computer Graph. Applic.* **13**, 48–56.

ZEVENBERGEN, L. W. and THORNE, C. R. (1987) Quantitative analysis of land surface topography. *Earth Surf. Process. Landforms* **12**, 47–56.

CHAPTER FIFTEEN

Dynamic maps for spatial science: a unified approach to cartographic visualization

JASON DYKES

15.1 INTRODUCTION

Maps have traditionally been employed both to store geographic information, and as a means of analysing spatial distributions and processes. The effort required to create a complex map manually has meant that single representations have been used for both objectives. Considerable research effort has gone into achieving cartographic products from which values can be read, and patterns detected, for numerous different variables. The advent of digital spatial information and computer graphics, mean that huge data sets which contain massive amounts of complex temporal and spatial behaviour are commonplace, and, new quick and dynamic means of displaying data are available. Despite this much of the endeavour in computer cartography has gone into digital and algorithmic reproductions of traditional, manual, cartography. Across the sciences researchers are increasingly relying on visual engagement with their data for analysis, yet the irregularity and volume of geographic data along with the alternative focus of computer cartography have meant that a unified environment for geographic visualization has not been forthcoming. This chapter describes the use of graphical user interface (GUI) widgets in computer cartography, and shows how they can form a unified environment which permits full cartographic visualization.

15.2 VISUALIZATION IN SCIENTIFIC COMPUTING (ViSC)

Developments in computer graphics and modelling have fuelled a recent propensity for visual representations of data in the natural sciences (McCormick *et al.*, 1987). Scientists frequently model and observe their data and measurements via computer simulations and screen graphics, using workstation based visualization systems. With such tools researchers can control model and display parameters and change their view of data in real time, yielding numerous concurrent views which may be dynamically linked. Information is gleaned from data by investigating, probing, transforming and re-displaying images on the computer screen. This method of analysis, termed 'visualization', or 'ViSC', provides at worst an overview of the data set, and at best insight from which new ideas about the data are developed and tested.

In the spatial sciences, technological advances have lead to a tendency for geographic data to be digital, and computer mapping is now standard. While there are still questions to be asked and papers to be written over the quest for a 'better' map, a new perspective exists which mirrors the ViSC trend in the natural sciences. It uses the tools available to the cartographer to produce multiple views of the same, or related, data in order to investigate spatial information. In addition, a fresh confidence is apparent in the use of images rather than numbers for describing geographical data. Numerous characteristics of ViSC are relevant to computer cartography and can be incorporated into computer maps. These include the ability to interrogate maps/images for information, the provision of numerous different 'views' of a data set, user control over cartographic and display parameters, the linking of views so that related information can be identified in each, and the availability of a temporal dimension for series and symbolism. If cartographers can embrace these features of ViSC and embellish them with their techniques an effective means of handling and analysing spatial data may be developed.

15.3 GIS FOR VISUALIZATION?

The complex, irregular nature of geographic data means that no integrated system exists which embraces the full range of visualization techniques available. Geographic information systems provide some of the capabilities found in ViSC systems: pseudo 3D viewing, point and click interrogation, control over classification, interactive colour schemes and multiple viewpoints are commonplace and more advanced features such as real time fly-bys, abstract data transformations, variation of the viewed variable in real time and some linking of views are becoming more widespread. However, the development of GIS from a quantitative setting has resulted in efforts being made to model spatial information in a complex database from which numbers can be extracted and statistics computed. Cartographic concerns usually involve the replication of traditional maps and the production of high-quality output of superb precision. Maps are built in as a static final product. The development of structures which permit the kind of real time dynamic display, visual interrogation, and engagement between researcher and data that characterize ViSC, have been secondary. This means that those who wish to employ aspects of ViSC in their analysis often have to use a GIS for geo-referencing and computing their data, before exporting to bespoke visualization software in order to animate, transform or fly by their data, or to link multiple maps and graphs of the data dynamically. Examples of such software are Macromind Director (see Weber, 1994), Dorling's (1993) cartogram algorithm, particularly the 'boot laces and sticky-tape' method of producing a fly-by from various sources (see MacEachren *et al.*, 1994), and the REGARD program of Haslett *et al.* (1991) or MacDougal's (1992) Polygon Explorer. There is currently no method of achieving each of these examples of visualization in a single piece of software or programming environment.

One way of resolving this problem is for smoother and closer links to be forged between software that manages and transforms spatial data and that which displays it. An alternative is to link display and data more fully in a data structure that revolves around graphic objects representing spatial entities and encapsulating spatial and attribute information. This chapter presents an example of the latter, by demonstrating a way in which some ViSC characteristics can be combined with cartographic techniques by mapping two dimensional spatial data as symbols in an object based graphic user interface environment.

15.4 Tcl/Tk – LINKING DISPLAY AND DATA IN A NON-CUMBERSOME WAY

Ousterhout's (1994) Unix based tool command language, 'Tcl', offers programming structures within a windowing environment, and 'Tk' is an X11 toolkit which defines familiar 'widgets' such as buttons, labels and scale bars. The two can be linked to build widgets that issue commands and arrange them on screen. Commands and widget behaviour can be coupled to particular mouse and cursor combinations. Such an environment, which permits interaction between screen and cursor, and screen objects, provides a basis upon which cartographic visualization can be developed.

The following section provides some simple examples of cartographic visualization using Tcl/Tk, showing how quickly and easily dynamic maps can be achieved. Each is accompanied by a figure containing a script in a window labelled 'Tcl/Tk Script' along with the widgets that result. Windows with titles that are numbered '#N' show changes in widget appearance. The code is stylized slightly, as the commands that place widgets in windows are omitted and some variable substitution is simplified in places. The full code, colour copies of the images, and example Tcl/Tk cartographic scripts are freely available from the URL referenced at the end of this chapter. Full appreciation of the techniques described requires hands on experience with the scripts. The examples emphasize the way in which cartographic visualization can be easily achieved by defining Tcl/Tk interface objects and providing values for the display options that they possess.

15.4.1 Widget definition and configuration

In Tcl/Tk widgets are created by stating a widget type and a name. Widget characteristics, or options, are then defined by declaring an option and its value. An example widget is shown in Figure 15.1. A button, '.b1' , is defined with options for text, width, height, background colour and command (to execute). These options have the values 'Show Time', 24 (characters), 3 (lines), Grey90 and 'exec date'. When clicked this button executes the Unix 'date' command. The values of widget options can be modified by issuing a 'configure' command. A re-configuration of the button '.b1' could change any option, for example colour from Grey90 to the darker Grey70, or text from 'Show Time' to 'New Title'. A simple example of widget communication is achieved by including the re-configuration of one widget in another widget's command option. In Figure 15.1 a second button, '.b2', has been defined so that when pressed, it changes the title and colour of the first button, '.b1', as suggested above. Figure 15.1 shows the buttons before (top) and after (bottom) the lower button has been clicked.

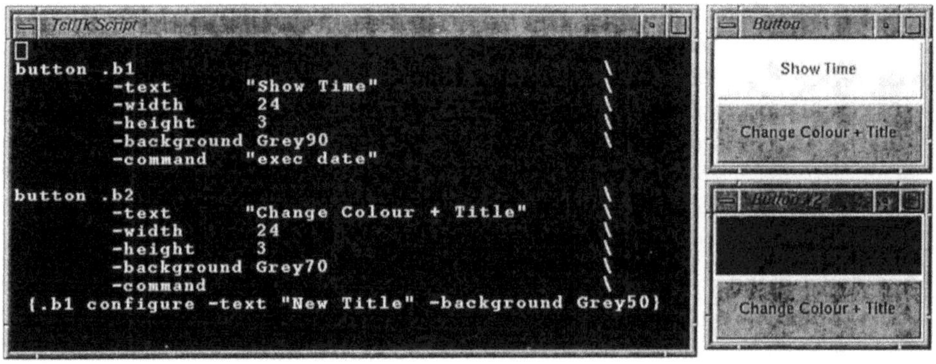

Figure 15.1 Button widgets with option-value pairs.

15.4.2 Basic cartographic symbolism

In addition to familiar GUI widgets, Tcl/Tk has a canvas widget which provides a co-ordinate plane in which graphic items such as lines, arcs, ovals, polygons, text and images can be located. Like button widgets, canvas items also have characteristics which are defined as option-value pairs. For example, line items have options for width, smoothing technique, bitmap fill and colour. This permits cartography as symbolism can be set to portray variable values, and dynamism as items can be re-configured or programmed to respond to other widgets or the cursor.

The following cartographic examples use a simple spatial data set comprising of four areas, or wards, for which boundaries are stored as lists of co-ordinate pairs in an array '*poly*' and centroids are held in arrays '*x*' and '*y*'. Three example attributes are retained in a four by three array '*val*' which is structured (area number, attribute number). The data used for the examples are depicted in Figure 15.2a. The reader should note that variables are set '*set variable value*' and variable values are expressed with a preceding '*$*' sign in Tcl/Tk.

```
# -- Array of lists of enclosing polygon coordinates.
  set poly(1) "100 100 100 195  50 195   5 150   5 100 100 100"
  set poly(2) "100 100 100 195 150 195 195 150 195 100 100 100"
  set poly(3) "100 100 100   5 150   5 195  50 195 100 100 100"
  set poly(4) "100 100 100   5  50   5   5  50   5 100 100 100"

# -- Array of attributes : value(ward,attribute).
  set val(1,1) 10 ; set val(1,2) 80 ; set val(1,3) 60
  set val(2,1) 40 ; set val(2,2) 70 ; set val(2,3) 90
  set val(3,1) 70 ; set val(3,2) 20 ; set val(3,3) 50
  set val(4,1) 80 ; set val(4,2) 50 ; set val(4,3) 30

# -- Arrays of centroid coordinates : x(ward), y(ward).
  set x(1) 50  ; set y(1) 150
  set x(2) 150 ; set y(2) 150
  set x(3) 150 ; set y(3) 50
  set x(4) 50  ; set y(4) 50
```

Figure 15.2a The simple spatial data set.

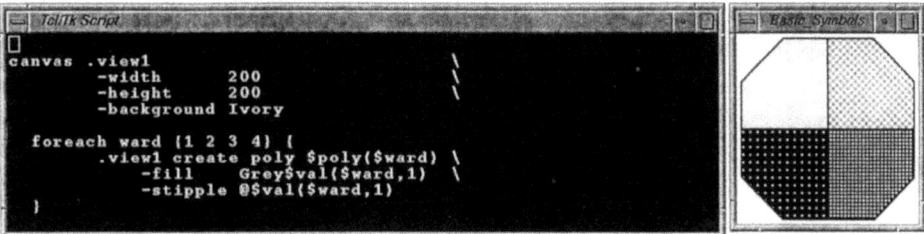

Figure 15.2b   Basic cartographic symbolism in Tcl/Tk.

A map is created in Figure 15.2b by first defining a canvas, '.*view1*'. Here the width, height and background colour options are given values. Commands of the form '*canvas-name create item-type coordinate-list [option value . . .]*' which plot an item or symbol in a canvas are then issued four times by a loop. Here polygons are created with co-ordinates from the array of lists '*poly*' depending on the looping variable '*$ward*'. Each symbol is given a shade (fill) and a texture (stipple) which represent the value of variable 1 for the ward in question, '*$val($ward,1)*'. The result is a simple choropleth map. When canvas items are created, each is given a unique integer ID, and may additionally be given a 'tag' comprising any string. Item tags and IDs can be used to select specific items for canvas commands, such as re-configuring.

15.4.3   Changing cartographic symbolism

As with buttons and other GUI widgets, canvas items can be re-configured in real time in response to other widget activity. Thus a polygon which forms part of a choropleth map can change colour or texture if it is reclassified by a scale bar, or if a new variable is to be plotted within it. This behaviour is demonstrated in Figure 15.3 which shows a loop creating three buttons, '.*b1*', '.*b2*' and '.*b3*', each of which passes the number '*$v*' (1, 2 or 3) to the procedure '*reconfig*'. The procedure uses the '*itemconfigure*' command which is the equivalent of '*configure*' for canvas items. The format is '*canvas-name itemconfigure ID/tag [option value . . .]*'. A loop ensures that each of the four zones is configured in turn and the values for the fill and stipple symbolism options are derived from the attribute array '*val*'. The three 'Reconfigure'

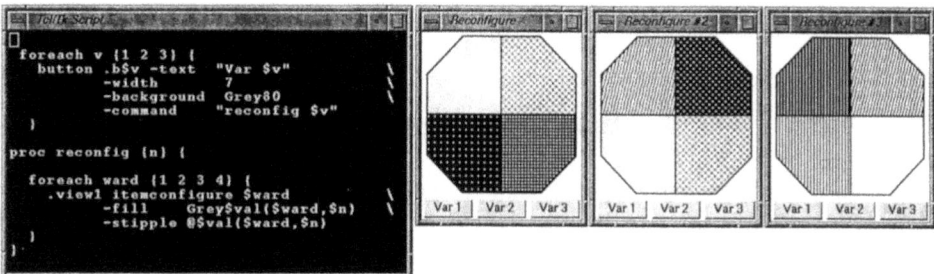

Figure 15.3   Changing cartographic symbolism with 'itemconfigure'.

windows in Figure 15.3 demonstrate that by pressing buttons 'Var 1', 'Var 2' and 'Var 3' the mapped variable is changed.

15.4.4   More complex symbolism and alternative views

Numerous different views of data can be created with these features. For example, scatter plots are symbols located by attribute rather than geographic space, and proportional circle maps locate circles geographically with radius and other symbol options depending on particular variables. These two views are shown in Figure 15.4, where a legend has been added to illustrate the full range of greys and textures, along with labels to indicate the variables being mapped and the symbolism being used in each view. Both maps create a new canvas, '.view2' and '.view3', and use the oval item type which is defined with a command of the form '*canvas-name create oval left top right bottom [option value . . .]*'. The circle map '.view2' symbolizes all three attributes by varying grey shade, texture and circle size. Up to eight variables could be mapped in a single symbol if oval height and width depicted different attributes, and the colour was defined as a red/green/blue (RGB), or hue/saturation/ value composite of three variables. The scatter plot '.view3' uses attributes 1 and 2 from the '*val*' array and a standard radius of 15 to locate symbols and attribute 3 for grey shading. Two lines are added in the scatter plot view to act as axes for reference.

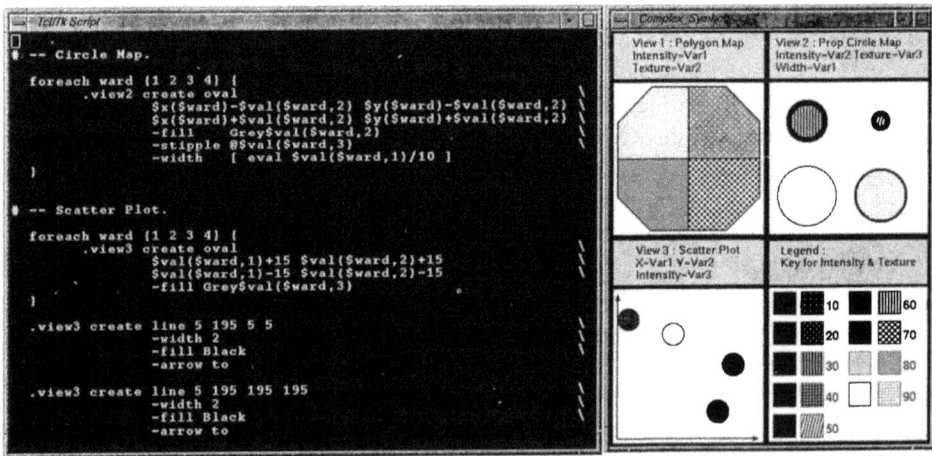

Figure 15.4   Complex symbolism and multiple views.

15.4.5   Interacting with maps

Interaction with canvas items is enabled by '*bind*' statements. These link mouse movement over a canvas to actions and can be used to interrogate maps for data as shown in Figure 15.5. Firstly a label widget, '*.l*', is specified. The '*-textvariable info*' option-value pair stipulates that the text displayed in the label is that held in the variable '*info*'. Whenever '*info*' is re-set, the text in the label will change accordingly.

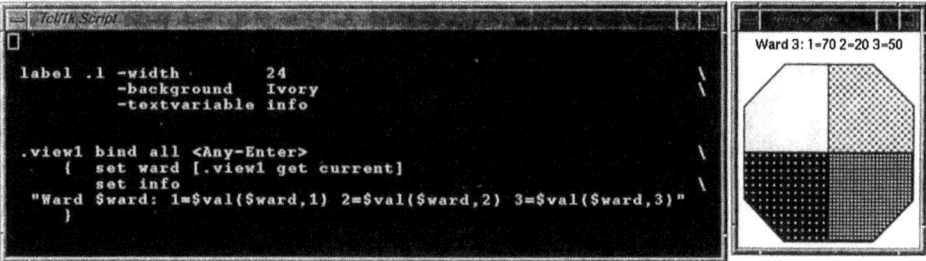

Figure 15.5 Interrogating symbols for values with 'bind'.

A '*bind*' statement of the form '*canvas-name bind ID/tag ⟨mouse-action⟩ {command}*' denotes that whenever any item is entered by the cursor the variable '*ward*' is set to the ID of that item. It also sets '*info*' to a text string containing information about the three attributes for that ward, which is consequently displayed above the map in the label '*.l*'. The result is a map that instantly displays information for any zone when the cursor is moved over it.

Binding can be used to perform more complex operations such as item reconfiguration. In Figure 15.5, 'all' items are bound to the interrogation commands. Groups of items with a common attribute, or tag, can be re-configured when any item with that tag is touched if tags other than 'all' are specified as conditions to bind commands. This type of binding is useful for investigating symbols which have common attributes, and is achievable between canvases or views.

15.4.6 Linking views

Maps or views can be linked by binding canvas items to a procedure which updates all items with a similar tag or ID across all views. This form of geographic brushing (MacDougal, 1992) is illustrated in Figure 15.6, where the first loop takes each canvas in turn, and binds cursor entry into all items to a series of commands. The first three commands find the ID of the ward that has been touched and re-set the value of '*info*' in a fashion demonstrated in Figure 15.5. The fourth command passes the ID of the current ward to the procedure '*update_views*'. This procedure has one argument and loops through every view configuring each item with the ID or tag that has been passed into it with a distinctive colour and texture. Zones on a choropleth map can be linked with points on a scatter plot, or circles on a cartogram in this manner. In Figure 15.6 the bottom right zone in the polygon map has been touched, and has consequently changed colour and texture. The symbols which represent the same ward in other views have also been re-configured as they have the same ID ('*\$ward*').

Tcl/Tk canvases also incorporate facilities for instant scrolling and scaling, movement of items, transformation, destroying items and raising or lowering items within a canvas. Additionally canvases can perform topological operations including finding the bounding box of an item, the closest item to a point, items which overlap a set of co-ordinates, or items within a certain distance of a point. These capabilities are of great use to the visualizing spatial scientist.

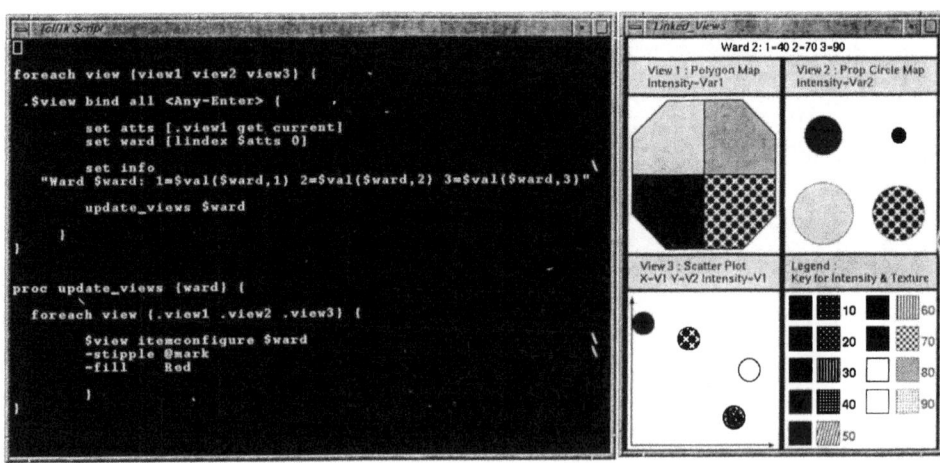

Figure 15.6 Linking multiple views with Tcl/Tk.

These modest examples illustrate a model for cartographic visualization using Tcl/Tk, and show the way in which a series of rudimentary commands can be used to create a sophisticated tool for visualizing spatial data.

15.5 CARTOGRAPHIC VISUALIZATION FOR COMPLEX SPATIAL DATA SETS

This approach, and these methods are being implemented on a larger scale in a cartographic data visualizer, 'cdv', which is being developed in Tcl/Tk to integrate ViSC techniques with computer cartography. The aim is to provide spatial scientists with a fully dynamic visual environment where questions can be asked and ideas developed. Two spatial data sets are used here to demonstrate the potential value of cartographic visualization. It should be noted that visualization is dynamic and interactive, and the static images provided with this chapter can only give an impression of the software and the success of the technique.

15.5.1 Viewing the geography of the UK census tables

UK census data and the 1981 ward boundaries for Leicestershire are displayed using cdv in Plate 4 (see colour plates section). This screen from a cdv session demonstrates ways in which the Tcl/Tk techniques outlined above can be used to visualize area data, by producing multiple, linked, interactive, cartographic views that can be interrogated, changed and manipulated in real time.

The focus of cdv is the 'Plot Variables' window (bottom centre) with three scale bars through which variables are selected. In Plate 4, Tables 4 and 10 of the UK Census have been loaded into Tcl arrays. Variables 335, 336 and 337 are selected; they represent proportions of total Irish-born (335), Irish-born males (336) and Irish born females (337). The main polygon map is initially shaded by the variable speci-fied by the top scale bar, as shown in Figure 15.2b. Shading changes as the bar is moved as demonstrated in Figure 15.3. Different views of these variables can be produced by creating polygon maps, circle maps and scatter plots, which can be coloured in numerous ways (see Figure 15.4). The scatter plots numbered 13, 14 and 15 (bottom left) show each combination of the three variables and are shaded by variable 335. Light values represent low proportional Irish population. In Plate 4, the polygon map (top left) has been re-shaded with an RGB colour composite of the three variables. Thus, lighter zones are more Irish, greener colours indicating male dominated zones, and more blue wards having a higher proportion of Irish females than males. Alternative variables and combinations of variables can be shaded instantly by moving the scale bars. The second polygon map (bottom right) shows Leicester city centre which has been selected through Tcl/Tk's instant scaling facility. Three other scatter plots (top right) show Irish population (335) against two social variables, namely persons without access to a bath or WC (391, plot 19) and those with no car (401, plot 20), and the social variables plotted against one another (plot 22). Each scatter plot is again shaded by Irish population.

The two proportional circle maps (top centre) show the social indicators, and are also intensity shaded by Irish population. Circles can be 'pushed' to the back of the canvas by clicking them to explore the detail of the pattern. Dorling's (1993) carto-gram algorithm has been implemented to repel overlapping circles and provide a social, rather than geometric, spatial view. All items can be interrogated by clicking with the cursor in the manner shown in Figure 15.6. The views are fully linked and dynamic. When an item is touched by the cursor in any view the equivalent ward is highlighted in the polygon map. Thus a scatter plot or cartogram can be traversed and spatial location referenced instantly. Double clicking an item results in items representing the same ward being coloured in every view, so clusters can be investi-gated. Note how the pink ward in the north-west of Leicestershire is highlighted in all the views. This is achieved in the way demonstrated in Figure 15.7. Plots 16 and 22, which depict three variables, two through location and one through colour, can be used to identify wards of particular interest, where the colours of close circles contrast. These can be located in the polygon map by touching the circles with the cursor, and other variables mapped to try and gain insight into the discrepancies.

15.5.2   A day at the park in Schwalm-Nette

Tourist behaviour in the Nettetal, a park located in Schwalm-Nette on the German border with the Netherlands, is depicted in Plate 5 (see colour plates section). Detailed locational and personal attribute data have been collected at hourly inter-vals for nearly one thousand individuals over a four day period. Tcl/Tk has been used to link the data to line (for route) and circle (for area) items. A scale bar determines the time, and the line and circle items are configured with size and colour combinations depending on the number of individuals located on/in the route/area at that time. The software has advantages over a time-series animation for investigation in that the user has total control over the speed and order of the sequence.

Researchers can also vary the display and selection characteristics as widgets have been created to vary the cartography, and to restrict inclusion to individuals with specific characteristics. Distinct sections of the population can then be visual-ized and assessed. For example, in Plate 5 the attribute widgets have been used to select and map the distribution of German and Dutch people, in cars or on bicycles, who are in groups and currently inhabiting their permanent places of residence, at 14:00 hours. By moving the scale bar, or clicking the attribute buttons the corresponding maps are displayed and time and attribute series can be produced in real time, and the data analysed. Clicking the 'Perm' and 'Temp' check buttons would immediately map people with the same population characteristics other than their residence, and the distribution of those in temporary accommodation would be revealed. Moving the 'Year_of_Birth' bar would re-express the data as an attribute series, depicting spatial behaviour of ordered age groups at a certain time.

Dynamically linked map views of different scales, allow location within the park and individuals' characteristics to be compared with the distribution of tourist points of origin on a national map. These factors, along with tourist social groupings, are the main concern of the research into barriers to tourist behaviour in parks near European national borders. Questions such as 'What are the behaviour patterns of locals in the park?', 'Where do wind surfers come from?', and 'When do people from the Venlo area use Nettetal?' can be investigated by visualizing spatial data in this manner. Work is proceeding to produce RGB colour composite maps to combine and contrast groups with selected attributes, and also to use the symbols to map individuals, rather than spatial features. This will produce an interactive 'movie' of spatial behaviour during a day in the park. Individuals could be mapped as items with shape, colour, texture, and outline width conveying selected personal attributes, much in the way that Chernhoff faces have been used in static maps to portray multi-variate data (Dorling, 1994).

15.6 CONCLUSION

Combining ViSC techniques with two dimensional cartography can provide a superb environment for investigating large and complex spatial data sets. Error or bias can be detected, an overview acquired, an understanding of the nature of the data developed, and eventually, analysis undertaken, results produced and knowledge gained. The Tcl/Tk GUI language supplies a unified, object based approach to visualization in computer cartography as demonstrated by the examples and the 'cdv' software. There is a map size overhead, maps of ten or twenty thousand lines will work, but without the instant changes associated with the generalized maps shown here. This environment not only equips the computer cartographer with more than adequate techniques for symbolism, and excellent facilities for dynamic mapping, unifying traditional cartographic techniques and those used in visualization, but also embeds them in a rich, open and active community of Tcl/Tk software developers who swap ideas and code over the Internet. These are excellent conditions for those wishing to visualize data, as code is accessible, and scripts are modular and procedural with elaborate variable substitution capabilities. This means that each cartographic visualizer can take advantage of general procedures which can be adjusted to account for the peculiarities and quirks of every real world database.

The public domain nature of the source code means that Tcl/Tk extensions are being produced constantly, and the potential for further cartographic development is good. Widgets can be created which account for projections and provide north arrows, scales and legends. Items can be programmed with option-value pairs that match cartographic visual variables exactly and use cartographic language more fully than at present. Interfaces can be produced that provide alternative symbolism techniques and guidance stemming from the cartographic literature, and depending on the data type. Cartographic items can be coded for streams, roads, towns and other map objects which have options for scale, generalization and other forms of symbolism and so encompass established cartographic theory. Furthermore, researchers are already producing three dimensional extensions to Tcl/Tk suggesting that the environment may be capable of incorporating some 3D surface, object and vector visualization in the future, and there is potential for a unified approach to geographic as well as cartographic visualization.

All of the Internet resources cited in this chapter, along with Tcl/Tk code, can be accessed and assessed by loading a World Wide Web page at URL: *http:// www.geog.le.ac.uk/argus/maps*

ACKNOWLEDGEMENTS

The HEFC JISC New Technologies Initiative for funding Project Argus, under which much of the work has been undertaken. Wim van der Knaap and Birgit Elands at the Centre for Recreation and Tourism Studies at the University of Wageningen, The Netherlands. Project Argus colleagues Professor David Unwin, Dr Peter Fisher, Jo Wood and Katherine Stynes deserve thanks for help and encouragement. 'This work is based on data provided with the support of the ESRC and JISC and uses boundary material which is copyright of the Crown and the Post Office.'

REFERENCES

DORLING, D. (1993) From computer cartography to spatial visualization: a new cartogram algorithm. *Auto Carto* **11**, 208–217.

DORLING, D. (1994) Cartograms for visualizing human geography. In HEARNSHAW, H. M. and UNWIN, D. J. eds, *Visualization in Geographical Information Systems*. Chichester: Wiley, pp. 85–102.

HASLETT, J., BRADLEY, R., CRAIG, P., UNWIN, A. and WILLS, G. (1991) Dynamic graphics for exploring spatial data with application to locating global and local anomalies. *Am. Statist.* **45**, 234–242.

MCCORMICK, B. H., DEFANTI, T. A. and BROWN, M. D. eds, (1987) Visualization in scientific computing. Special issue ACM SIGGRAPH *Computer Graphics* **21(6)**.

MACDOUGAL, E. B. (1992) Exploratory analysis, dynamic statistical visualization, and geographic information systems. *Cartogr. GIS* **19**, 237–246.

MACEACHREN, A., BISHOP, I., DYKES, J., DORLING, D. and GATRELL, A. (1994) Introduction to advances in visualizing spatial data. In HEARNSHAW, H. M. and UNWIN, D. J. eds, *Visualization in Geographical Information Systems*. Chichester: Wiley, pp. 51–59.

OUSTERHOUT, J. K. (1994) *Tcl and the Tk toolkit*. Wokingham: Addison-Wesley.

WEBER, C. R. (1994) Multimedia authoring in macromind director. In MACEACHREN, A. M. and FRASER-TAYLOR, D. R eds, *Visualization in Modern Cartography*. Oxford: Pergamon, pp. 97–101.

PART FIVE

Applications

Precision farming in the office: investigating the development of a GIS-based field information management system

ALISTAIR GEDDES

16.1 INTRODUCTION

Arable farmers are beginning to use yield mapping systems to study the spatial variation in crop performance within their fields. Such systems are based on field data collected using Global Positioning System (GPS) receivers integrated with grain flow meters on combine harvesters. Similarly, GPS-based control systems have also been developed for chemical sprayers and fertiliser spreaders, allowing the application of variable dosage rates according to field treatment maps. Both types of system are being used commercially by farmers in the USA and Western Europe, including a small but growing number of growers in the United Kingdom.

Somewhat less attention as yet has been devoted to investigating the 'middle link', that is, to the requirements for integrated office-based field information management systems which would enable the storage and manipulation of yield maps, as well as the generation of treatment control maps for guiding future field operations. In principle, the capabilities of GIS technology for handling spatial data could provide the basis for such systems.

At the same time, however, the reality must be faced that a number of difficulties persist in the collection of spatially variable yield data from combine harvesters. The occurrence of errors in the raw data, particularly those associated with the accuracy of sampling position locations, may have serious implications for the reliability of generated yield maps and thereafter for the quality of any subsequent GIS-derived information. The large sizes of the raw harvester files add a further difficulty. Suitable methods must be found for addressing these errors during the data logging and/or processing stage(s).

In this chapter, a data-processing sequence based on the kriging method of spatial interpolation is presented as a possible GIS 'work flow' for deriving yield maps. The possibilities for extending this work flow to incorporate the integration of yield maps as well as other field data is also briefly discussed.

16.2 HARVESTER DATA LOGGING REGIME

A variety of techniques for recording the spatial variability in arable crop yields has been considered in the literature over the past decade (Stafford *et al.*, 1991; Stafford and Ambler, 1991, 1992). In the UK, the weight of interest has centred on the development of a commercial yield mapping system (Massey Ferguson, 1993) linking GPS position measurements with yield readings from a grain flow sensor on board combine harvesters.

16.2.1 Harvester positioning within fields using GPS

Global positioning systems (GPS) make use of earth satellites and associated ground control stations to allow suitable receivers to determine their location on the Earth. Greatest attention has focused on the NAVSTAR GPS system owned by the US Department of Defence. The components of this system are described fully elsewhere (see Wells *et al.*, 1986). Civilian access to the NAVSTAR satellite constellation since 1983 has encouraged the rapid uptake of GPS as an increasingly common method for collecting field data.

GPS relies on 'satellite ranging' whereby position is calculated within the receiver by measuring its distance from a group of satellites as a function of the travel time of the signals from these satellites (Cumings, 1994). With measurements to a minimum of three satellites, receiver position can be calculated using triangulation techniques; measurement to a fourth satellite is also required to solve for clock error relating to the atmospheric delay of the signals (Figure 16.1).

Positioning accuracy relies on the method of distance measurement employed by GPS receivers. Civilian receivers, including the harvester-based system, observe the Coarse Acquisition (C/A) timing codes of the satellite signals, known as 'pseudo ranges'. Such measurements are susceptible to various errors, including satellite drift, internal clock errors, and atmospheric interference and reflection of the satellite signals. Experiments have shown that the accuracy possible with stand-alone civilian pseudo-ranging receivers is in the range of 15 m to 20 m. However, the Department of Defence reserves the right to downgrade the C/A timing code accuracy through a cryptographic technique termed 'Selective Availability' (SA). When SA is operational, the Department has stated that 95 per cent of GPS-derived positions will fall within 100 m of their true locations (August *et al.*, 1994).

In the context of UK agriculture, results derived using a single receiver alone when SA is active are not sufficient for positioning moving machinery within arable fields (Stafford and Ambler, 1991; Macleod, 1992a). To improve on this, the commercial yield mapping system uses two receivers working in tandem, a technique known as differential GPS (DGPS). DGPS exploits the assumptions that (a) measurement errors are largely independent of the receivers, and (b) similar errors are common to all receivers observing the same set of satellites under the same conditions over a wide area. In order to gauge these errors, the antenna of a reference receiver must be placed at a point of accurately known location. Corrections to the

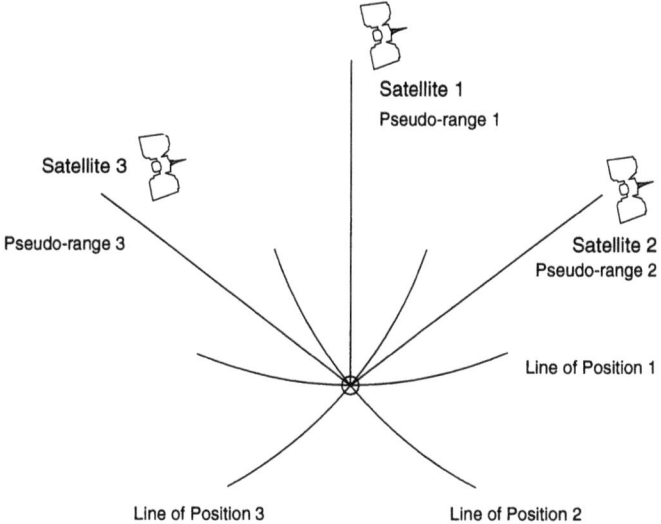

(Note: in practice, the three pseudo-ranges do not converge at a single point. They are adjusted in equal

amounts until the resulting Lines of Position converge. The amount of error is the clock error and the

point of convergence is the position.)

Figure 16.1   Determining GPS receiver position using pseudo-range measurements.

observed pseudo-ranges and range rates may then be calculated, and are subsequently applied to fixes made by the second 'roving' receiver.

In the commercial yield mapping system, the mobile unit antenna is mounted on the combine harvester cab, whilst the reference station is fixed on the roof of either a farm building, or at the manufacturer's local dealership. At the time of investigation, pseudo-ranges were measured using six channel receivers, each channel tracking a single satellite during cutting operations. Differential positioning was performed in real time by the harvester receiver, using correction data broadcast via a radio link from the reference station by a 0.5 W or 10 W transmitter. (It is worth noting that yield mapping does not require position data to effect instantaneous control over machine operations; hence these corrections could equally have been post-processed). Latitude and longitude co-ordinates are logged in the WGS84 (World Geodetic System 1984) datum. Using this DGPS set-up, the manufacturers reported five metres horizontal accuracy as the normal working limit (Moore, 1994).

16.2.2   Yield meter measurements

Grain flow meters have been introduced by a number of combine harvester manufacturers over the last decade. Their purpose is to provide estimates of harvested crop yield, both as 'spot' tonnes per hectare (tha^{-1}) readings during cutting and as a summary of total harvest weight for a field. Various meters have been tried in yield mapping experiments, although at the time of writing, the system based on the

'Flowcontrol' meter (Massey Ferguson, 1993) was the only one commercially available to UK growers which allowed continuous recording of yield levels during harvesting.

Yield determination with the 'Flowcontrol' meter works on the principle of constant flow velocity of material movement across a calibrated sensor. This is approximately the case for small grain crops when the sensor is mounted on an elevator discharging clean grain into a holding tank inside the harvester. The instrument can thus be calibrated in terms of mass flow rate. Grain mass is measured by the attenuation of radiation between a small gamma source and detectors. Harvester forward speed and cutting width data are used to calculate a spot reading for grain mass per unit area over a two second period (Murphy *et al.*, 1994).

Measurement of grain mass per unit time is the principal strength of the 'Flowcontrol' meter. Other systems which instead use volumetric measurements are more susceptible to errors introduced by changes in the type of crop under harvest and grain moisture content. This advantage has been demonstrated in field trials comparing calibrated weight estimates from the meter against weighed yields. Results for cereals indicate an accuracy of greater than 2 per cent (Stafford *et al.*, 1991).

16.2.3   Harvester positioning accuracy: requirements versus capability

For successful yield mapping, it is essential that data logging preserves the true fluctuations in yield levels as the harvester progresses along each cutting swathe. Correlating DGPS position fixes with 'spot' data generated by the yield meter provides the basis for this task. Yet data collection is problematic.

First, assessing the accuracy of DGPS techniques for the agricultural situation is open to two questions of fundamental importance: (a) how much spatial variability actually exists in yield levels, and (b) how much spatial variability can be identified using DGPS techniques? Finding answers to these questions is no easy task. It is clear from the literature that significant variability in soil and crop properties may exist within fields (Stafford and Ambler, 1991; Macleod, 1992a). Equally apparent is that the level of variability itself differs, depending on the property being measured as well as between fields, farms and regions. Hence the specification of a definitive set of positioning accuracy requirements for arable operations is extremely difficult. Working in consultation with the UK Agriculture and Food Research Council, Macleod (1992b) suggested a base positioning specification for field mapping operations of plus or minus 0.5 m over ranges up to 500 m. At this level of accuracy, the implication is that current DGPS techniques for locating combine harvesters are lower than Macleod's specification by one order of magnitude.

Second, maintaining DGPS mode across an entire field demands that both receivers observe the same satellites throughout cutting operations. However, the surrounding terrain and presence of obstructions such as woodland and pylons within and around fields may raise problems of signal loss and distortion. Additional difficulties may result from the constantly changing nature of the 'crossing angles' formed between a receiver and the four orbiting satellites which it is using. Generally, more accurate positions can be resolved when these angles are large. This effect of geometry is normally calculated by receivers as a ratio known as the Dilution of Precision (DOP), which is multiplied by the pseudo-range errors to give an indication of overall positioning accuracy. When DOP values exceed some pre-set threshold, receivers may stop processing positions for the current satellite set.

Ideally therefore, advance DGPS 'mission planning' should be performed to minimise these degrading effects. In pre-planning, additional software may be used to select the best period for a future data collection session. Satellites to be tracked should be chosen based on knowledge of their current positions and availability, and by predicting and testing suitable DOP thresholds. Equally, users should be aware of the fundamental accuracy of DGPS-derived data and potential error sources. Unfortunately, the relative novelty of the technology militates against this. Results from a telephone questionnaire survey of 13 farmers using the 'Flowcontrol'-based yield mapping system during the 1994 harvest indicated that these users were wholly unaware of the requirement for pre-planning (Geddes, 1994). Only three felt able to offer numerical estimates of the limits of DGPS for positioning their harvesters.

16.2.4   Yield meter calibration

Achieving accurate yield measurements is dependent on correct calibration of the yield meter, both prior to and during cutting. The accumulation of crop residue and dirt with the harvester mechanism during continuous operation is inevitable, particularly with oilseed and protein crops such as rape and peas. When such a build up occurs on the surfaces of the yield sensor, the effect is an artificial increase in the level of yield recorded. The operator must periodically check and, if necessary, re-calibrate the meter.

16.2.5   Difficulties with integrating DGPS data and yield data

Harvester operating conditions may vary considerably within a field, according to the type and health of the crop, soil and topographic conditions, and the occurrence of pest damage and weeds. These factors introduce contingent difficulties for the correct correlation of GPS fixes with measurements from the yield sensor.

Reporting on this situation, Murphy *et al.* (1994) divided these errors into (a) systematic errors concentrated at the head of each cutting strip, due to a lag time between the start of cutting and the arrival of grain at the flow meter, and (b) errors distributed about the field, associated with sudden disturbances in normal operation.

Figure 16.2 illustrates the effect of lag time on averaged yield readings recorded at the beginning of four harvest swathes in a single crop. Over a period of approximately 40 seconds at the start of each swathe, readings form a curve rising to a point of inflection where 'normal' grain flow across the yield meter is established. All data recorded within this initial period are potentially erroneous. Moreover, it is likely that the duration of this lag period will be peculiar to each harvester and crop.

The second error source relates to disturbances in normal harvester operations. Causes of disturbances are essentially unpredictable (stones becoming trapped in the cutting header for instance), although some warning signs may be detectable (where, say, crops lie flattened). Their effect can lead to sudden fluctuations in forward speed

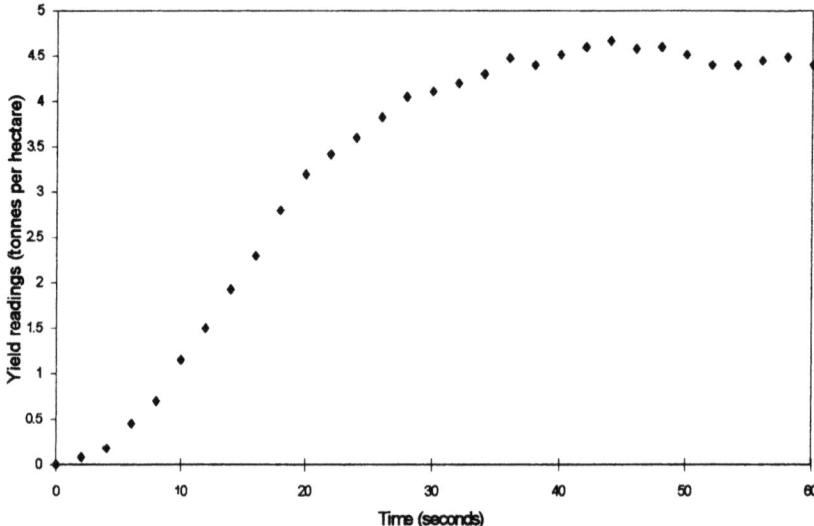

Figure 16.2   Mean yield readings over initial 60 s of a harvest swathe (after Murphy *et al.*, 1994).

and manoeuvres in which the harvester is operating but not cutting. Co-ordinates with very high associated yield readings may thus indicate periods of rapid deceleration when greater crop masses per unit time were being processed, and vice versa.

16.3   SPATIAL INTERPOLATION OF YIELD MAPS

Data logging on the harvester results in a digital file containing integrated position/yield readings. Following file transfer to a farm office PC (using a solid state memory card), yield maps may be created by interpolating continuous surfaces from these data.

The harvester-based system differs in two key respects from conventional geostatistical sampling approaches. First, yield data are recorded at far greater densities than would usually be considered; based on an average harvester speed of 8 to 10 km h^{-1}, data densities may be anywhere between 300 to 800 readings per hectare. Second, the accuracy of the input readings can be highly variable, resulting from the difficulties outlined above. The interpolation algorithm must be sufficiently robust to address this error variability while preserving the 'true' inherent spatial variability within the recorded yield data. In light of these requirements, the kriging method of interpolation was investigated as a means for generating yield maps.

Three main advantages for using kriging are offered in the literature. Firstly, properties of the natural environment are commonly observed to vary randomly across space. Kriging is based on the principle that the property of interest varies continuously in space with random and spatially dependent components of variation. As such, Oliver and Webster (1990) report that the kriging modelling method is more suited to fitting this reality than the simple mathematical models of other interpolators. Secondly, variance estimates can be determined for each of the predicted values from which an interpolated map is constructed. These may also be mapped, thereby allowing an assessment of the confidence limits which can be associated with the interpolated results. Thirdly, kriging uses a moving elliptical neighbourhood search function to compute estimates. Prior structural analysis of the spatial dependence in the data is used to control the size and orientation of this ellipse, and to assign weights to the observations falling within it. The method can thereby be tailored to the size of the particular area under study (Rogowski, 1995). Thus in principle at least, kriging can provide an efficient approach for interpolating yield maps.

ASCII files containing harvester data were acquired for a 15.2 ha test field in Bedfordshire for successive crops of winter wheat in 1992 and 1993. Data logging was as described previously. Unfortunately no information was given regarding the precise location of the reference antenna, or whether the same harvester was used on both occasions. Data were processed following the work flow outlined in Figure 16.3.

16.3.1 Preliminary inspection of harvester data

Differences in the DGPS recordings between the two seasons are illustrated by the scatter plots in Figure 16.4, using the first 800 fixes recorded in both files. These fixes correspond to the headland areas which were harvested first. Examination of these plots illustrates the variability inherent in receiver accuracy. A number of outliers

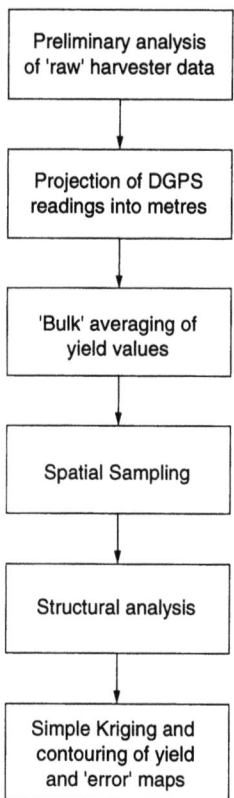

Figure 16.3 Work flow for processing harvester data.

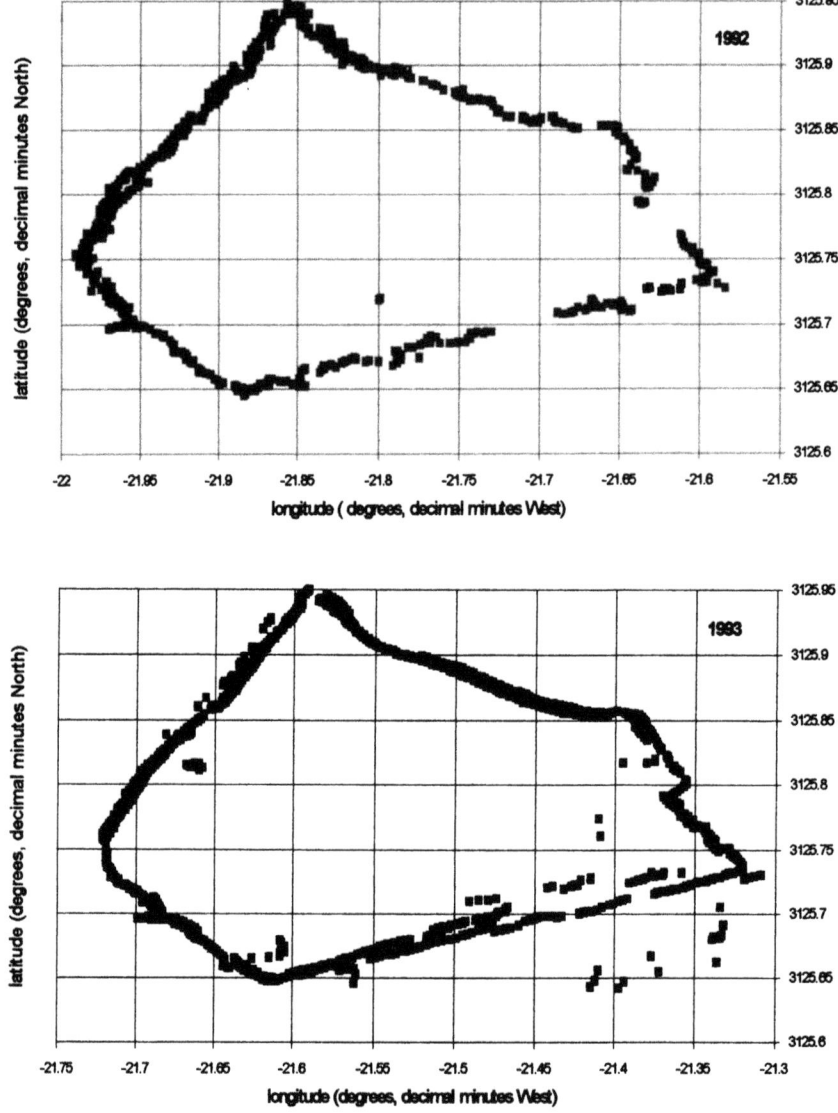

Figure 16.4 Distributions of initial DGPS fixes.

are separated from the majority of fixes in both cases, especially in the 1993 data. Directional biases were evaluated simply by comparing the position fixes recorded in each season for four vertices which were identified along the field boundary. The mean deviation in longitude and latitude values suggested an eastern shifting of approximately 20 m in the 1993 data.

16.3.2 Projection, 'bulk' averaging and sampling

In soil and land resource surveys, bulking is a statistical technique commonly adopted to improve the precision of measurements for a property A from n samples drawn within a local area. Bulk averaging acts to 'smooth' out the influence of extreme outliers in a data set. Averaging was similarly applied to the projected data prior to their interpolation because of the many erroneous readings that may have been recorded. To carry this out, the data files were first projected from spherical co-ordinates into metres (in single precision) using the Great Britain National Grid option supported in ARC/INFO.

Mean yield values were calculated for each position fix, based on the neighbouring values lying within a 10 m radius. Although this choice of radius size was somewhat arbitrary, the key feature to note is that it was greater than the 5 m accuracy limit reported for the DGPS receiver. Figure 16.5 illustrates the frequency histograms for the 1992 data set. Since winter wheat yields of 10 tha^{-1} are exceptional, the lower graph presents a much more feasible set of values.

In effect, the original yield readings associated with the point fixes are replaced with values representing the average yield level within a 314 m^2 area centred over each fix. By so expanding the zone of influence around each point, the total number of sampling points needed to cover an entire field is reduced. Consequently, statistics may be used to draw samples of lower densities from the data files.

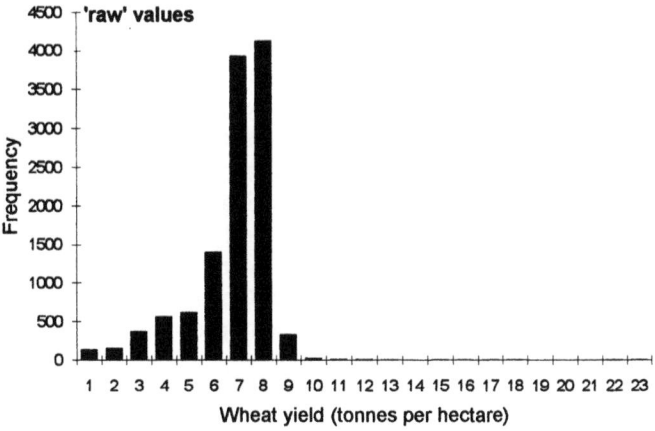

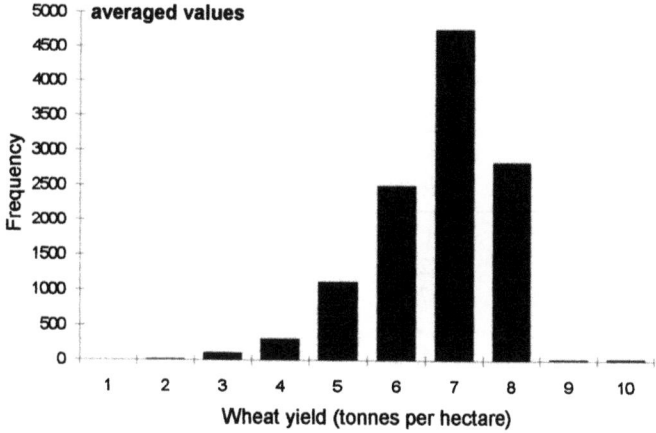

Figure 16.5 Frequency histograms for 1992 yield data.

Some guidance in this matter was obtained by consulting a previous soil fertility study (Webster and McBratney, 1987). Subsequently, a comparable sampling interval was used to draw a sample from the 1992 file. In order to provide a contrast, a contrasting sample was drawn using a much smaller interval, giving a total size close to the maximum limit for the kriging software (GEO-EAS, version 1.1; Englund and Sparks, 1988). Systematic sampling was achieved by first sorting the data in terms of increasing magnitude of x and y co-ordinates.

Table 16.1 summarises the statistics for the two samples, as well as for the original and bulked yield files. The average yield of 6.3 tha^{-1} is consistent across all data sets. None of the distributions illustrated skew to a degree which required normal transformation.

16.3.4 Structural analysis

Applying regionalised variable theory for yield mapping assumes that the variation in a yield value A at a point x can be expressed as the sum of three major components:

$$A(x) = \mu + \varepsilon'(x) + \varepsilon''(x) \tag{16.1}$$

where μ represents a structural component associated with a constant local mean yield, $\varepsilon'(x)$ is a locally varying, spatially correlated component, and $\varepsilon''(x)$ denotes spatially uncorrelated residual 'noise' (Burrough, 1986). In mapping yield patterns across a field, the spatially dependent component represented by $\varepsilon'(x)$ is of greatest interest. Variation in this element between yield measurements may be summarised by a function known as the variogram. The variogram describes how the variance of paired sample measurements $\{1/2 \, VAR \, [A(x) - A(x + \mathbf{h})]\}$ changes with the separation distance. Typically all possible sample pairs are examined and grouped by separation vectors ('lags') $\mathbf{h}$ of equal distance. On this basis, they can provide estimates for suitable weighting coefficients to use in predicting yield values from neighbouring observations.

Scatter plots for the 1992 samples showed that the influence of harvester cutting direction was evident in the distribution of values both around the field perimeter and in a strip pattern following the cutting swathes. The existence of these patterns meant that the experimental variograms could not be considered solely as a function

Table 16.1 Summary statistics for study yield data

	Original yield data	'Bulked' yield data	Sample one	Samples two
Sample size	11 724	11 724	101	972
Mean (tha^1)	6.26	6.26	6.26	6.26
Median (tha^{-1})	6.7	6.5	6.3	6.5
Min, max (tha^{-1})	0, 22	1.2, 9.9	2.9, 7.8	2.1, 9.6
Standard deviation	1.65	1.04	0.91	1.02
Variance	2.735	1.075	0.826	1.035
Skew	−0.63	−1.0	−0.85	−0.96
Kurtosis	6.61	1.43	1.09	1.19

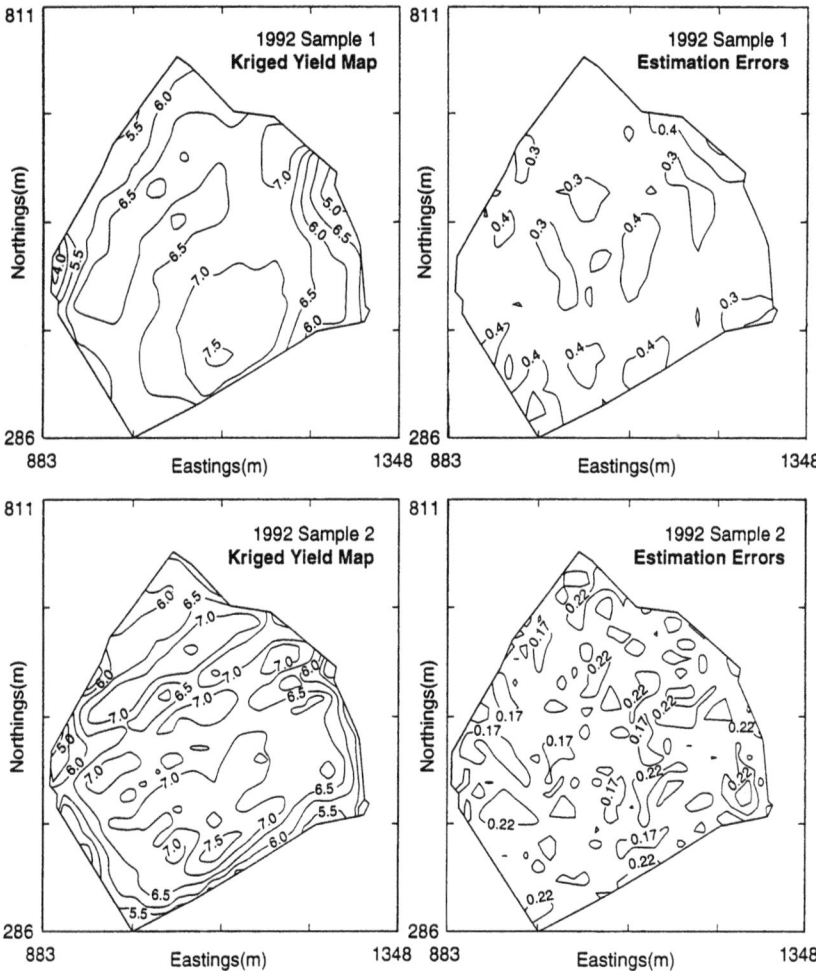

Figure 16.6 Contour plots for kriged yield estimates and estimation errors.

of the distance between paired observations. Information on anisotropy – directional bias – was obtained by computing directional variograms for the samples. Subsequently, spherical models were fitted using the accepted least squares method (van Deursen and Wesseling, 1993).

16.3.5 Kriging and contouring of yield estimates

The fitted variogram model parameters were used to derive weights for interpolating yield map estimates. Assuming that the mean value μ in equation 16.1 can be accurately determined, the method of kriging known as simple kriging gives best linear predictions (Pebesma, 1995). The constant average figure of 6.3 tha^{-1} obtained across Table 16.1 indicates that this is approximately the case with the yield samples. This mean value is used to ensure that the weights sum to one, thereby ensuring a lack of bias.

To make a map, estimates must be kriged at regularly spaced intervals. For the purposes of this study, estimates were kriged at intervals of 15 m, roughly 10 per cent greater than the adopted lag separation. Slightly elliptical neighbourhood search ellipses were defined with the major axis in both cases lying across the harvester cutting direction. Map estimates were subsequently kriged using a block size of 60 m.

Contouring of the predicted estimates and their standard errors produced the field maps for the 1992 winter wheat crop shown in Figure 16.6. Generally the trend of spatial variation is the same for both samples, although the sample pattern is smoother for the smaller sample. Yields above 7 tha^{-1} occurred in the south of the field. From this area, levels fell off gradually towards the northern and western boundaries, although a finger of slightly poorer performing ground (below 6.5 tha^{-1}) intruded into this pattern from the south-west. Below average crop performance was concentrated around the perimeter, where the impact of vehicles turning on the headlands was evident.

The low contour values on the 'error' maps indicated that the kriging standard deviation (and hence the variance) is almost uniformly low over the field. Values are slightly higher for the smaller sample, since the kriged estimates in this case were interpolated from a more sparse data set. Since the kriging standard deviation is dependent on the variogram model used, these errors should be interpreted with caution. However, Webster and McBratney (1987) suggest that the degree of this model-related error is not serious when, as in this study, large samples are used to compute the variogram.

16.4 DISCUSSION AND CONCLUSIONS

Questionnaire results suggest that only on very large winter wheat holdings in the UK are there the resources to investigate yield mapping. Furthermore, the findings also indicate that the farmers who have adopted this technology are currently using annual yield maps as simple visual guides, most notably for helping to tackle problems associated with soil compaction. However, such decisions are reactive.

The ability to make proactive judgements instead is fundamental to current proposals for using GIS for farm-based management. If stable spatial patterns of crop performance can be identified from a digital fields database, potentially they could be drawn on to help make a range of projections, including the production of treatment control maps and forecasting likely profit margins for new crops and management strategies. In this respect, Schnug *et al.* (1993) have advanced what amounts to a raster-based overlay model for identifying areas of consistently similar relative crop performance within fields, termed 'equifertiles'. For each cell subdivision on a field map, an equifertiles value representing aggregate yield performance could be obtained by summing the values for the corresponding cells from yield maps generated in each season. Relative rather than absolute values should be used, since the range in arable yield values for a field may vary from year to year, depending on the crop type and prevailing growing conditions.

If forecasts derived from equifertiles are to carry weight in the farmer's decision making process, then the model inputs representing the annual yield maps must be reliable. However, this cannot be assumed. Considerable difficulties exist in integrating variable DGPS measurements with yield readings on board operating harvesters. Repeatability of position measurements is a fundamental criterion for consistent yield mapping from one season to the next. Equally, the possibilities for using DGPS-controlled field treatment maps to target inputs relies on the capability for accurately re-locating different parts of a field. Yet the comparison of the 1992 and 1993 headland plots conforms with results from other experiments (August *et al.*, 1994), which have suggested that considerable variation in DGPS-derived positions may exist for any given location from one time period to the next.

Kriging is a rigorous interpolator which would allow quantitative information about yield mapping errors to be incorporated directly within the farmer's decisions. Such information potentially provides a means for tailoring the basic equifertiles model to match the quality of the input maps (for instance, adding appropriate weighting coefficients). A number of other kriging trials have indicated that variogram ranges for arable crop yields consistently exceed 50 m and frequently are greater than 100 m (Murphy *et al.*, 1994). Together with an equifertiles map, knowledge of this spatial dependence may help in establishing suitably spaced points for soil sampling across a field. Moreover, if harvested yield levels are assumed to serve as valuable indicators of biological factors affecting crop performance, then this correlation could be exploited by co-kriging analysed soil samples with the more dense yield data.

Comparison of the map results for the 1992 samples indicates that sampling selection involves striking a balance between the level of detail that is desired and the level of prediction error that is acceptable. By careful sampling, the quantities of data recorded in a field may be reduced to allow more efficient processing without reducing the quality of the map results. Considering the large file sizes generated by the harvester system, data sampling could become an important requirement for reducing the load on farm computers following yield mapping over several seasons.

Methods for kriging and incorporating errors have received attention in GIS research and development only relatively recently. Whether they can, and will, be integrated as effective tools within a GIS-based field information management system remains a challenge for future research.

ACKNOWLEDGEMENTS

This chapter gives a summary of dissertation research while on the Edinburgh University MSc GIS course in 1993–94. The contributions of M. Moore of Massey Ferguson for the test data, P. Massam, and T. Malthus of the Geography Department are gratefully acknowledged.

REFERENCES

AUGUST, P., MICHAUD, J., LABASH, C. and SMITH, C. (1994) GPS for environmental applications: accuracy and precision of locational data. *Photogram. Eng. Remote Sensing* **60**, 41–45.

BURROUGH, P. A. (1986) *Principles of Geographical Information Systems for Land Resources Assessment.* Oxford: Clarendon Press.

CUMINGS, D. A. B. (1994) Positioning geography: an appraisal of the Global Positioning System (GPS). Unpublished MSc dissertation, University of Edinburgh.

ENGLUND, E. and SPARKS, A. (1988) *GEO-EAS (Geostatistical Environmental Assessment Software) – User's Guide.* Las Vegas: U.S. Environmental Protection Agency.

GEDDES, A. (1994) Precision farming in the office: investigating the development of a GIS-based field information management system. Unpublished MSc dissertation, University of Edinburgh.

MACLEOD, F. (1992a) The physical and conceptual exploitation of the inherent spatial variability of the land. Unpublished PhD thesis, University of East London.

MACLEOD, F. (1992b) Agricultural vehicle positioning and its integration with a large-scale land information system *J. Navigation* **44**, 30–36.

MASSEY FERGUSON (1993) *Yield Mapping System.* Stareton: Massey Ferguson Group Ltd.

MOORE, M. (1994) Personal communication.

MURPHY, D. P. SCHNUG, E. and HANEKLAUS, S. (1994) Yield mapping – a guide to improved techniques and strategies. Proceedings of the 2nd International Conference on Site-Specific Management for Agricultural Systems, University of Minneapolis, March 1994.

OLIVER, M. A. and WEBSTER, R. (1990) Kriging: a method of interpolation for geographical information systems. *Int. J. Geogr. Inform. Syst.,* **4**, 313–332.

PEBESMA, E. J. (1995) *GSTAT User Manual.* Bilthoven: National Institute for Public Health and Environmental Protection.

ROGOWSKI, A. S. (1995) Quantifying soil variability in GIS applications. I. Estimates of position, *Int. J. Geogr. Inform. Syst.,* **9**, 81–94.

SCHNUG, E., MURPHY, D. P., HANEKLAUS, S. and EVANS, E. J. (1993) Local resource management in computer aided farming: a new approach for sustainable agriculture. In FRAGOSO, M. A. C. and VAN BEUSICHEM, M. L. eds, *Optimisation of Plant Nutrition.* Netherlands: Kluwer Academic Publishers, pp. 657–663.

STAFFORD, J. V. and AMBLER, B. (1991) Dynamic location for spatially selective field operations. Presentation at the 1991 International Winter Meeting of the American Society of Agricultural Engineers, Chicago, December 1991.

STAFFORD, J. V. and AMBLER, B. (1992) Mapping grain yield variation for spatially selective field operations. Presentation at the Agricultural Engineering International Conference, Uppsala, June 1992.

STAFFORD, J. V., AMBLER, B. and SMITH, M. P. (1991) Sensing and mapping grain yield variation. Presentation Proceedings of the Symposium 'Automated Agriculture for the 21st Century' of the American Society of Agricultural Engineers, Chicago, December 1991, 356–365.

VAN DEURSEN, W. P. and WESSELING, C. G. (1993) *The PC-raster Package.* University of Utrecht, Department of Physical Geography.

WEBSTER, R. and MCBRATNEY, B. (1987) Mapping soil fertility at Broom's Barn by simple kriging. *J. Sci. Food Agric.* **38**, 97–115.

WELLS, D. E., BECK, N., DELIKARAOGLOU, D., KLEUSBERG, A., KRAKIWSKY, E. J., LACHAPELLE, G., LANGLEY, R. B., NAKIBOGLU, M., SCHWARZ, K. P., TRANQUILLA, J. M. and VANICEK P. (1986) *Guide to GPS Positioning.* Fredericton, N. B., Canada: Canadian GPS Associates.

CHAPTER SEVENTEEN

Modelling air quality and the effects on health in a GIS framework

CHRISTINE E. DUNN and SIMON P. KINGHAM

Establishing associations between environmental variables and ill health poses a number of significant challenges. The present chapter proposes a strategy for utilising GIS as a framework for integrating air quality and health datasets in an attempt to define more clearly the nature of health–environment relationships. We assess the value of incorporating air quality models into GIS in an examination of the links between outdoor air quality and reported respiratory ill health.

17.1 INTRODUCTION

Methodologies for assessing the impact of airborne pollution on health status are generally poorly developed, although there is growing concern about indoor and outdoor air pollutants (including, most recently, road traffic emissions) as a potential source of chronic and acute adverse health effects. The relationship between air quality and ill health in a population is complicated not only by the underlying presence of confounding variables such as lifestyle, occupation and socio-economic factors, but also by the nature of exposure which invariably involves a complex mix of compounds. Examination of this issue therefore calls for a multidisciplinary approach, requiring the expertise of epidemiologists, toxicologists, statisticians and social scientists. In this respect, medical geographers and others have turned to Geographical Information Systems (GIS) as an integrative tool for analysing health–environment relationships, and have achieved varying degrees of success.

Previous studies using GIS and spatial analysis to assess the links between air quality and health have often made assumptions about spatial variations in air quality (Diggle *et al.*, 1990; Gatrell and Dunn, 1995). Thus, distance decay approaches have been applied where proximity of disease cases (often referenced by postcode of home address) to a putative source of atmospheric pollution (e.g. heavy industrial complexes, single sources such as incinerators or nuclear power plants, road traffic) is used as a proxy for personal exposure. Much of this work has been carried out without recourse to air quality data as a means of substantiating hypotheses about total risk and exposure. This may be because GIS users, and, in particular geographers, lack the expertise and access to technology to enable, for example, primary air quality monitoring programmes to be undertaken. Routine air quality monitoring, which in the UK is incorporated under the remit of local authorities, provides an alternative data source, although data may not be available for the relevant pollutants, at the required resolution or time intervals, or for appropriate locations. Attempts to provide true measures of total air quality in any case suffer from a number of difficulties. In particular, levels of 'traditional' atmospheric pollutants such as smoke and sulphur dioxide are of less significance today than during the first half of this century and other contaminants, notably VOCs (volatile organic compounds), oxides of nitrogen, PAHs (polycyclic aromatic hydrocarbons), particulates and ozone are now of more concern. The complex nature of some of these 'newer' pollutants makes it difficult not only to isolate single pollutant species for analysis, but also to consider the potential impact for health of interaction effects. Establishing theories about causal mechanisms is yet more problematic and here GIS is likely to play only a small role.

All air quality monitoring data, regardless of the quality and resolution of the sampling framework, are limited by the fact that the data pertain to fixed points in space. In studies relating health status to atmospheric variables, therefore, it is important to qualify the representativeness of such point data in terms of wider areal units, since an air quality sampling framework will almost always consist of fewer points in space than are available for health datasets. The use of interpolation techniques to address this issue has undoubtedly been facilitated by the inclusion of algorithms in many GIS packages, notably Thiessen polygons, trend surface analysis and kriging.

A valuable alternative to assessing spatial and temporal patterns of air quality, especially when suitable measured data are unavailable, is the use of air quality models. Such models represent a means whereby 'emissions can be related to atmospheric pollutant concentrations' (Seinfeld, 1986, p. 600) and entail using the characteristics of a specific pollutant source(s) in conjunction with ambient atmospheric parameters, notably meteorology, in order to predict the spatial behaviour and dispersal of a pollutant. Air quality modelling has been widely used for predicting spatial patterns of pollutant dispersal. However, although research integrating environmental modelling within GIS is evident (Goodchild et al., 1993), that utilising modelled air quality data is less well documented, especially in relation to health. Fedra (1993) combined an air quality modelling approach with GIS for simulating environmental impacts and developing planning strategies, while Collins et al. (1995) combined dispersion modelling, statistical interpolation and rule-based techniques in a GIS-based approach for deriving estimates of exposure to linear sources of nitrogen dioxide (road traffic). The latter work forms part of a larger study looking at links between traffic pollution and respiratory health in children (Briggs et al., 1994). The present chapter considers the value of incorporating air quality modelling into a GIS framework (using ARC/INFO) to produce a 'surface' of pollutant isolines for a set of compounds (total VOCs) from a stationary source. Combining air quality estimates with primary survey data for health status in adults and children, we explore spatial variations in respiratory ill health in terms of quality of outdoor air to assess whether emissions from a specific industrial source may be influencing health status. We now describe the background to the study site for which the approach was tested.

17.2 STUDY AREA

Public awareness of the potential impact of environmental factors on health is often raised in those living in close proximity to industrial complexes (the so-called 'skyline effect'). The present study arose from one such concern expressed by residents living near to a wallpaper factory in County Durham, UK. The study site is semi-rural and much of the population is concentrated in a single housing estate within 2 km to the north-east of the factory site. There are no other major sources of industrial pollution in the vicinity. Public complaints centred around the nature of the smell of airborne emissions from the factory and gave rise to an earlier investigation of asthma prevalence using repeat prescription data for anti-asthma medication from local GPs (Dunn *et al.*, 1995). The earlier work used GIS to define potential areas of exposure in the community as simple sectors of circles of increasing distance from the factory, and demonstrated a 24 per cent excess of anti-asthma prescribing for the population living in the area within 1 km to the north-east (downwind) of the factory. Further revision of areal sectors revealed that the increased prevalence was restricted to middle-aged and elderly residents living between 500 m and 1000 m to the north-east of the factory. In a move towards a more realistic definition of outdoor pollutant exposure, the present study, which forms part of a larger, multidisciplinary investigation of health status, and socio-economic and environmental factors (Kingham *et al.*, 1995) uses GIS to integrate modelled air quality with health-related information for the local population.

17.3 DATA SOURCES

17.3.1 Health survey

To provide basic information on health status, housing and lifestyle for both adults and children we used a community survey (self-completion postal questionnaire) with a total sample size of 4500 and achieving a response rate of 65.3 per cent for adults and 55.8 per cent for children. Conversion of residents' unit postcodes to OS grid references in order to provide a spatial point reference for each respondent can be achieved by means of postcode matching, using the Central Postcode Directory, CPD (OPCS, 1985). In view of the small size of the study area, however, and the limited accuracy of the CPD (100 m) it was necessary here to allocate unique grid references to each address for which a questionnaire was completed. This was carried out manually by using large scale OS maps to extract grid references accurate to the nearest 10 m. A set of attribute variables for each case in the health survey was then appended to the spatial dataset after importing to ARC/INFO.

17.3.2 Air quality modelling

Several different types of mathematical air quality model are available and they vary in terms of data requirements, accuracy and cost. One of the simplest and most widely used models is the Gaussian plume model which assumes that the distribution of pollutants in the horizontal and vertical directions is Gaussian in shape. This model is characterised by constant source strength, constant wind speed and direction, constant diffusion characteristics and non-reacting pollutants (Guldman and Schefer, 1980). Some of the better known and more readily available Gaussian models are those produced by the United States Environmental Protection Agency (USEPA) and included amongst these are the Industrial Source Complex (ISC) models as used here. These can calculate short-term or long-term pollutant concentrations and can account for multiple stack sources by summing the concentrations predicted from each individual source, but cannot take into account any synergistic properties of combinations of pollutants. Two sets of input data are required; stack source data and meteorological data. The source data consist of location, height, temperature and diameter of the stack(s), and gas exit velocity and emission rate. For the site of interest here, this information was provided by the factory management for all four chimney stacks. The meteorological data comprise date, hour (GMT), wind direction, wind speed, ambient temperature, stability class and mixing height. Although in the UK several meteorological stations collect some of this information, only a limited number collect all the variables required, especially those of stability class and mixing height. In order to run the model, therefore, data generally need to be purchased from the Meteorological Office and in the case of this study, data were acquired for the nearest station for the five-year period September 1989–September 1994.

The model adopts a keyword/parameter approach to specifying options and input data. Thus, control options allow local land-use conditions to be specified in general terms on the basis of categories 'rural' or 'urban'. In the present study, a combination of rural and urban conditions was specified by classifying sectors of a circle (degrees from north) on the basis of land-use information derived from secondary map sources and from local knowledge of the area. The model also provides flexibility in terms of receptor sites for which predicted pollutant values are required. In this case, model receptors were located at ground-level and based on x,y grid points on a circular polar network centred on the factory stacks. Receptors were placed at seven downwind distances, to a maximum of 2 km and at every 10 degrees, giving a total of 252 receptors. The model was run using the complete set of meteorological data to predict a 5-year average of VOC concentrations in the study area.

17.4 ANALYSIS AND RESULTS

Concentrations of VOCs predicted by the air quality model for the polar receptor points were imported to ARC/INFO. It is then straightforward to use ARCTIN to convert these values to a TIN (Triangulated Irregular Network), and TINCONTOUR to convert the TIN to a surface of isolines, at intervals specified by the user. Figure 17.1 shows the output from this process for VOC isolines at 25 $\mu g/m^3$ intervals. Relatively high predicted concentrations of VOCs are predicted for the northern part of the study area and to the east of the factory, reflecting the effects of prevailing winds. The contours clearly demonstrate the impact of plume dispersal in the immediate area of the factory in that the highest concentrations are found, not immediately downwind of the source, but at some distance away. Highest concentrations (of up to 174 $\mu g/m^3$) are reached at 100 m east-north-east of the factory while levels are generally reduced to 5 $\mu g/m^3$ at a distance of about 2 km from the factory.

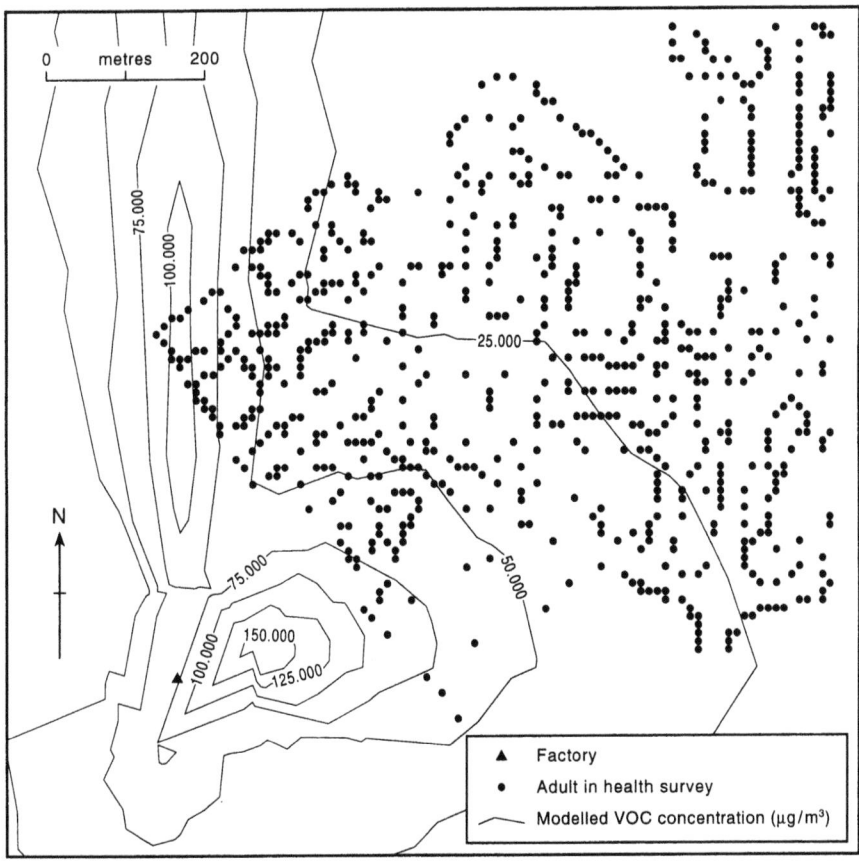

Figure 17.1 Adult questionnaire respondents and modelled levels of VOCs.

These interpolated VOC levels then formed the basis of an analysis to test whether spatial variations in air quality were associated with respiratory symptoms reported by respondents to the questionnaire survey. Focusing on community questionnaire returns for reported asthma, hay fever and bronchitis, contoured values of VOCs were overlaid onto point data sets pertaining to reported ill health for these conditions in both children and adults. This then allowed us to perform appropriate statistical analyses. First, odds ratios were used to measure the association of exposure and morbidity. The odds ratio is the ratio of the chance of exposure among the cases of a disease to the chance in favour of exposure among a set of controls (Beaglehole *et al.*, 1993). These ratios were calculated from contingency tables defining 'high risk' and 'low risk' areas in the community based on separate predicted levels of VOCs. Areas with greater than 25 μg/m^3 VOCs and greater than 50 μg/m^3 VOCs were selected. Within ARC/INFO it was possible to define areas bounded by these modelled pollution levels and then to ascertain the number of cases of the chosen measure of ill health within the 'polluted' or higher risk area. For asthma, cases and controls were defined by their answers to two questions relating to asthma in the health survey; these were: 'have you ever had, or been told that you have had asthma?' and, asked later in the questionnaire, 'have you ever suffered from asthma?'. The second question was included as a check and, in most, but not all cases, responses to both questions were the same. The number of cases within the

Table 17.1 Odds ratios for respiratory ill health and modelled VOC exposure areas

Measure of ill health	Modelled VOC concentration ($\mu g/m^3$)	Odds ratio	95% Confidence limits
Children			
Asthma ever suffered	25	1.45	0.94–2.24
Asthma ever suffered	50	1.56	0.71–3.41
Asthma ever had	25	1.27	0.83–1.95
Asthma ever had	50	1.31	0.59–2.92
Adults			
Asthma ever suffered	25	0.91	0.67–1.23
Asthma ever suffered	50	0.82	0.52–1.31
Hay fever	25	0.73	0.49–1.09
Bronchitis	50	1.03	0.63–1.68

higher risk ('polluted') area is compared with an expected number based on the hypothesis that the pollution has no effect on the distribution of people suffering from the various forms of ill health (Kelsey *et al.*, 1986, p. 107).

Odds ratios were therefore used to ascertain whether there was a statistically significant difference between the expected and observed numbers of cases of reported symptoms within the 'polluted' area. A value of 1 indicates that the observed number of cases is as expected. A value below 1 indicates that there are fewer than the expected number of cases in the 'polluted' area, and a value above 1 indicates that there are more cases in the 'polluted' area than expected. Odds ratios for selected measures of ill health in relation to exposure to VOCs are shown in Table 17.1. Some of the odds ratios, particularly those for children's asthma show results above 1, suggesting that the number of cases of children's asthma in the area designated as 'polluted' is greater than expected and that cases are more likely than the controls to have been exposed to relatively high levels of VOCs. From 95 per cent confidence limits, however, (Kelsey *et al.*, 1986, p. 109), none of the health variables examined in relation to the modelled pollution showed any statistically significant association.

In order further to explore the evidence for the effect on respiratory health of exposure to VOCs, we carried out logistic regression analyses using selected health-related variables from the community survey as dependent variables, and modelled VOC concentrations, and socio-economic and housing factors (occupation, overcrowding, cigarette smoking, dampness in the home and type of domestic heating) as explanatory variables. In ARC/INFO a predicted VOC concentration was 'attached' to each point (home address) in the data sets for questionnaire respondents, treating adults and children separately. This was carried out firstly by using the TINLATTICE command to convert the TIN described earlier to a lattice file. LATTICESPOT was then used to interpolate VOC values from the lattice, for each point in the point coverage. Regression results demonstrated that in all but one case, the reporting of long-term illness, modelled VOC levels had no statistically significant association with reported health outcomes.

17.5 DISCUSSION

Relationships between environmental exposure and ill health are highly complex and in order to achieve a full understanding of the processes and causal mechanisms involved, an integrated and multifaceted approach is called for. Analysis of health data in relation to environmental variables in a GIS framework goes some way towards helping us to understand these relationships and we have demonstrated here the application of one useful technique for clarifying the role of ambient air quality in determining respiratory health status. For a case study involving self-reported symptoms in residents living in close proximity to a wallpaper factory which emits volatile organic compounds (VOCs), analyses produced no positive associations with the health survey data. There appears in this case, therefore, to be no significant association between estimated VOC levels and respiratory ill health. A number of factors need to be considered in reviewing this conclusion, however.

First, predicted VOC concentrations represent only a proxy for environmental contamination and degree of exposure. Actual total personal exposure may vary between individuals as a result of occupation, lifestyle, daily activities and movements, and individual susceptibility; furthermore, some of these risk factors may interact. The mechanisms and effects of such interactions are extremely difficult fully to address and may call for more sophisticated monitoring and analysis in terms of personal or biological monitoring, approaches which are beyond the scope of the present research. Some attempt at extricating the impact of risk factors on health has been made here by regression analysis.

Second, the nature of the relationship between exposure and response varies depending on the mechanism by which the pollutant causes disease. For most inhaled pollutants, evidence indicates increasing risk with increasing exposure and a linear relationship without a threshold is assumed (Kingham *et al.*, 1995). The pollutants examined here, VOCs, form a group of compounds and ideally the analysis should be extended to look at more specific constituents.

Third, the success of the proposed approach rests to a large extent on the reliability of the air quality model used. The Gaussian USEPA ISC model used here makes a number of assumptions about chemical interactions, distribution and stability of constituents in the plume, and constancy of wind speed and direction (Lyons and Scott, 1990; Hewitt and Allott, 1992). In particular, no account is taken of other sources of emissions, including road traffic or fugitive escapes from the factory itself (transient emissions from sources other than the main stacks). The model therefore predicts concentrations for a factory 'baseline' only. Other errors can occur as a result of inaccurate meteorological and emissions data; in the present case modelling was necessarily based on data from a meteorological station several miles away from the study area. Field measurements of VOCs carried out at a limited number of sites in the local environment indicated that, using daily meteorological data, the model was inefficient at predicting *absolute* concentrations for individual days (Kingham *et al.*, 1995). Using the longer-term meteorological data, however, *relative* predicted concentrations match expectations both in terms of prevailing wind conditions and with respect to spatial patterns of VOCs measured in the community. More complex air quality models are available but these suffer from more stringent requirements in terms of data input and, in practical terms, may be prohibitive in terms of cost (Sawford and Ross, 1985). For an overview of relative variations in air quality over geographical space, therefore, an approach combining simple modelling with the capabilities of GIS may be appropriate.

In conclusion, air quality modelling clearly depends not only on the accuracy of the algorithms employed, but also on the quality of the input data; it is not a direct substitute for carrying out an environmental monitoring programme. Where the latter presents difficulties or is inappropriate, however, whether due to cost or lack of expertise and technology, modelling in conjunction with GIS may provide one way of testing and developing hypotheses about health–environment relationships. We suggest that this approach may offer a way forward for medical geographers and others in attempting to extend their application of GIS as a tool for understanding spatial patterns of respiratory disease.

ACKNOWLEDGEMENTS

This study is part of a multidisciplinary venture carried out by the authors in collaboration with the following: Professor Raj Bhopal, Professor Peter Blain, Mr Chris Foy and Dr Tanja Pless-Mulloli of the University of Newcastle upon Tyne, Dr Sushma Acquilla of Durham Health Commission and Dr Jane Halpin of the Northern and Yorkshire Regional Health Authority.

REFERENCES

BEAGLEHOLE, R., BONITA, R. and KJELLSTROM, T. (1993) *Basic Epidemiology.* Geneva: World Health Organization.

BRIGGS, D. J., ELLIOTT, P. and LEBRET, E. (1994) Analysing small area variations in air quality and health: the EC-SAVIAH study. Paper presented to the 6th International Medical Geography Symposium, Vancouver, Canada 12–16 July.

COLLINS, S., SMALLBONE, K. and BRIGGS, D. (1995) A GIS approach to modelling small area variations in air pollution within a complex urban environment. In FISHER, P. ed., *Innovations in GIS 2.* London: Taylor & Francis, pp. 245–253.

DIGGLE, P. J., GATRELL, A. C. and LOVETT, A. A. (1990) Modelling the prevalence of cancer of the larynx in part of Lancashire: a new methodology for spatial epidemiology. In THOMAS, R. W. ed., *Spatial Epidemiology.* London: Pion, pp. 35–47.

DUNN, C. E., WOODHOUSE, J., BHOPAL, R. S. and ACQUILLA, S.D. (1995) Asthma and factory emissions in northern England: addressing public concerns by combining geographical and epidemiological methods. *J. Epidem. Commun. Health* **49**, 395–400.

FEDRA, K. (1993) Clean air: air quality modelling and management. *Map. Aware. GIS Eur.* **7**, 24–27.

GATRELL, A. C. and DUNN, C. E. (1995) Geographical Information Systems and spatial epidemiology: modelling the possible association between cancer of the larynx and incineration in north-west England. In DE LEPPER, M. J. C., SCHOLTEN, H. J. and STERN, R. M. eds, *The Added Value of Geographical Information Systems in Public and Environmental Health.* Dordrecht: Kluwer Academic, pp. 215–235.

GOODCHILD, M. F., PARKS, B. O. and STEYAERT, L. T. (eds) (1993) *Environmental Modelling with GIS.* Oxford: Oxford University Press.

GULDMAN, J.-M. and SCHEFER, D. (1980) *Industrial Location and Air Quality Control: a Planning Approach.* Chichester: Wiley.

HEWITT, C. N. and ALLOTT, R. (1992) Environmental monitoring strategies. In HARRISON, R. M. ed., *Understanding our Environment*, 2nd edn. Cambridge: Royal Society of Chemistry, pp. 189–243.

KELSEY, J. L., THOMPSON, W. D. and EVANS, A. S. (1986) *Methods in Observational Epidemiology*, Oxford: Oxford University Press.

KINGHAM, S. P., ACQUILLA, S. D., DUNN, C. E., HALPIN, J. E., FOY, C. J. W., BHOPAL, R. S., BLAIN, P. and PLESS-MULLOLI, T. (1995) Health in the vicinity of industry in Bishop Auckland. Unpublished report, University of Newcastle upon Tyne.

LYONS, T. and SCOTT, B. (1990), *Principles of Air Pollution Meteorology.* London: Belhaven Press.

OPCS (OFFICE OF POPULATION CENSUSES AND SURVEYS) (1985) *Central Postcode Directory User Guide,* 2nd edn. Titchfield, Hampshire: OPCS.

SAWFORD, B. L. and ROSS, D. G. (1985) Workshop on regulatory air quality modelling in Australia, 8th International Clean Air Conference, *Clean Air,* August, 82–87.

SEINFELD, J. H. (1986) *Atmospheric Chemistry and Physics of Air Pollution.* London: Wiley.

CHAPTER EIGHTEEN

Land-user intentions and land-use modelling

ROBERT MacFARLANE

18.1 INTRODUCTION

The research on which this chapter is based has the objective of making a link between the application of GIS in supporting land-use policy decisions and a developing body of literature on farm response to external change, primarily contributed by agricultural economics and rural sociology. The primary research interest was in modelling the consequences of agricultural and rural land-use policy. As around 70 per cent of the UK is farmed in one form or another, agriculture is of considerable interest to geographers, not only as land-use, but also as a highly complex and diverse economic and social system.

GIS applications in supporting land-use policy decisions, further to the provision of spatially referenced inventories, seem to fall into one of two categories:

1. Modelling the impact of changes in the bio-physical environment on the parameters of agricultural activity and hence land-use patterns.
2. Modelling the response of farming regions and individual production units to changes in the economic and policy context.

The role and usefulness of spatially referenced inventories of land-use is well-documented and this chapter focuses on certain shortcomings of the latter two approaches. Examples of these approaches are often presented as 'models of land-use change', which it is suggested is a misrepresentation. The central thesis of this chapter is that it is not possible to accurately predict how individual parcels of land, and therefore the aggregate land-use pattern, will respond to changes in the external environment, without representing the conditioning effect of individual decision-makers. The primary interest of this research is in farm response to policy shifts, rather than response to more direct change in the bio-physical production environment such as climate shifts.

18.2 GEOGRAPHICAL (IS) PARADIGMS

It is argued that the two categories of 'land-use models' identified above are based upon very different, but equally sweeping generalisations about the nature of human decision-making. These are related to paradigms which are now increasingly regarded as out-of-date and somewhat inadequate in the context of geography as an academic discipline. Different perspectives on land-use adjustment have attempted to throw light on the endogenously, and exogenously induced stimuli for change. Three principal perspectives are discernible. The first is bio-physical determinism, whereby the physical environment is held to control decision-making; the second is economic determinism where farmers as producers react in a rational manner to prevalent economic conditions, in order to maximise profit; and the third stresses the rôle of individual socio-behavioural influences on the decision-making process (Tarrant, 1974). Taylor and Johnston (1995) identify three approaches to geographical understanding and explanation, suggesting that 'people seeking to meet the challenges of a rapidly-changing world are as likely to base their arguments on other philosophies as they are to embrace the empiricist-cum-positivist stance taken, at least implicitly, by most GIS advocates' (p. 55). The most relevant of these, in this context, are:

1. Empiricist–positivist geography and technical control equate prediction with explanation . . . Whereas some geographers believe that only the physical world can be treated in such a way, others argue that both individuals and the societies they create are open to similar 'science-based' manipulation . . .
2. Humanistic science and mutual understanding promote appreciation of events involving humans by uncovering and communicating the thoughts behind the action . . . (p. 55–56).

In paradigmatic terms, this research represents a shift from the first stance, which is underpinned by a deterministic conception of human behaviour, towards the second. The concern, however, is not simply one of 'academic' interest, but has real implications for much of the policy-support research concerning rural land-use, societies and economies.

Fundamentally, modelling is concerned with the abstraction and generalisation of the critical features of a given system (Macmillan, 1989). Geographers and other researchers using GIS as a tool to study and predict environmental change at a variety of scales have responded to the relative abundance of climatic, pedological, geological, biological and hydrological data, as compared to the paucity of spatially referenced socio-economic data, by limiting their models to an environmentally deterministic representation of land-use change.

Environmental determinism, a theoretical position according to which the nature of human activity is controlled by the parameters of the physical world, can be traced back to the work of Charles Darwin. The core ideas are those of natural selection and adaptation; the physical environment imposes controls to which humans can only respond. If land-use modelling is based on land capability data and normative decision rules which describe the interaction of individual farms, as purely productive economic units, with their environment then it essentially conforms to this conception of human-environment interaction. In practical terms, previous GIS-based models have been structured in such a way that the interaction of economic and bio-physical environmental factors are themselves *held* to affect individual farms. Underpinning this deterministic framework is the presumption of universal human rationality in making land-use decisions. Human behaviour is held as a constant.

18.3 TOWARDS THE REPRESENTATION OF COMPLEX DECISIONS IN COMPLEX ENVIRONMENTS

The concept of rationality is not a simple one. Economists and behavioural scientists differ widely in their application of the term. Economic research into land-use change has been based upon normative models of farm decision-making, founded on neo-classical economic principles. Rationality in decision-making is equated with the drive for profit maximisation. The modelling of land-use change in response to shifts in the bio-physical production environment is conceptually and paradigmatically very similar. The focus of a number of projects (for instance Aspinall, 1991) has been to relate shifts in the parameters of land capability to quantifiable changes in the economic margins of a range of productive enterprises, and hence predict a change in land-use patterns. Both assume that, individually and collectively, farmers and other land managers will react uniformly and 'rationally' to the changes in their external environment. Research into the behaviour and motivation of farmers and other 'economic actors' has shown this to be an unreasonable assumption (Barnes, 1988).

It all hinges on the concept of human rationality. Previous models of land-use response to economic and bio-physical shifts have overlooked the 'behavioural filter' of the proprietary land-unit. Such models have held this filter to be a constant – clear glass as it were – assuming that all affected individuals will re-evaluate their position and adjust their behaviour to maximise their profits. This overriding assumption, reducing all farmers to the caricature *homo economicus*, which populated so much early geographical location theory, is indeed pragmatic. All models are inescapably generalised, but this definition of rationality is increasingly untenable. *Homo economicus* has survived the rigours of paradigmatic shifts in geography (Johnston, 1983), and more recently in economics (*The Economist*, 24.12.94), and continues to 'drive' deterministic models of land-use change.

It is argued here that assumptions which reduce human rationality to the pursuit of a single and over-riding economic objective are relatively valid at the regional

Figure 18.1 Homo economix? (Originally appeared in *The Economist*, and reproduced with kind permission of the cartoonist, GED Melling.)

scale, providing aggregate estimates of land-use shifts and the macro-economic implications of policy change. The best example of such a normative geographical land-use model is the Land-Use Allocation Model (LUAM). This was initially developed to predict the regional consequences of shifts in the Common Agricultural Policy (Centre for Agricultural Strategy, 1986), and later refined and integrated with a GIS to provide predictions of land-use change at a greater range of scales (White, 1990). Harvey (1992, p. 26) argues that:

> Policy makers require some simplified picture or model of the complex interaction and relationships in order to analyse the consequences of policy options and thus choose the most appropriate policy set. If academics and professionals cannot or do not provide such models, then policy makers will use their own implicit, ill-defined and inconsistent mental pictures instead.

The justification for developing models of farm and land-use response to policy change is clear and undisputed, but the scope and structure of GIS-based models need to be extended to incorporate a wider range of factors which are known to condition the process of land-use change.

> Explanations of landscape change are complicated by the fact that policy, institutional and technological factors affect farmers first and pieces of countryside only second (Potter, 1986, p. 193).

Policy developments over recent years add further urgency to this requirement for GIS researchers and practitioners to incorporate human behaviour (within constraints imposed by the social, economic, political and physical context) as a spatial surface into GIS analyses and models. Commentators have traced a shift in emphasis from agricultural policy, to a more diverse rural policy agenda (Gilg, 1991). Concerns over surplus production, environmental degradation and the cost of agricultural support have provided the impetus for schemes to divert land out of agricultural production and de-couple production-related payments from financial support to farmers. Production maximisation has been displaced as the dominant objective of rural land-use policy, and farmers are increasingly required to consider a wide range of potential non-agricultural sources of income to maintain their livelihood. Under this revised policy agenda farm adjustment and land-use change can only be predicted with any degree of accuracy by representing the intentions, constraints and resources of individual landowners and land-users. Such an approach requires a 'bottom-up' approach to the representation of farms, and farming and land-use systems.

If new policy schemes are likely to be oriented towards specific land-management objectives, and be more sensitive to regional and local spatial variations in land capability, amenity value and conservation worth, for instance, then the rationale for policy-support research to reflect the ability and willingness of individual farmers to participate in those schemes is clear. Normative models of human behaviour cannot adequately represent the diversity of circumstances and characteristics which describe the farming community. GIS techniques are not, however, ruled out by a methodology which is focused on the internal dynamics of individual farms, and their interaction with the external policy-economic and bio-physical environment.

The research on which this chapter is based represents an attempt to bridge research gaps between geographical modelling of environmental change, conceptual

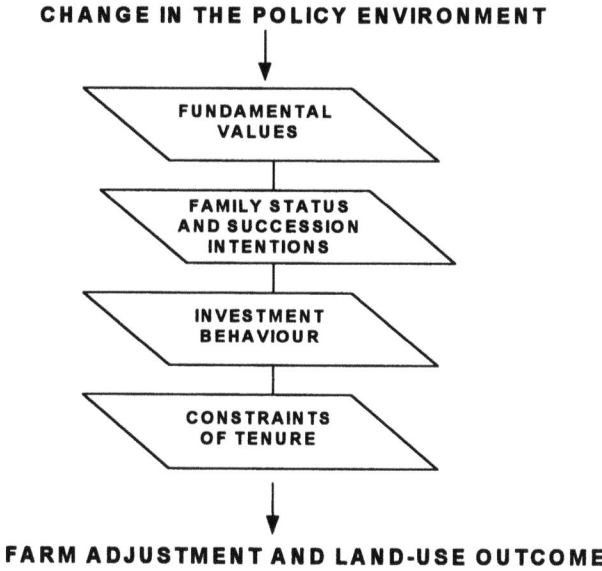

CHANGE IN THE POLICY ENVIRONMENT

FUNDAMENTAL VALUES

FAMILY STATUS AND SUCCESSION INTENTIONS

INVESTMENT BEHAVIOUR

CONSTRAINTS OF TENURE

FARM ADJUSTMENT AND LAND-USE OUTCOME

Figure 18.2 Socio-behavioural and economic filters conditioning farm adjustment.

and empirical research into farmer motivation and behaviour and aspatial methodologies which have been developed to represent the interaction of economic and socio-behavioural characteristics in predicting farm response to new policy measures and land-use schemes (Potter and Gasson, 1988; Brotherton, 1988, 1992).

For many farmers, their occupation is a way of life and pressure to develop 'alternative' enterprises such as tree crops or tourism will be strongly resisted. Figure 18.2 illustrates farm response to change in the external environment in terms of a set of socio-economic and behavioural 'filters'. Research has examined the effect of many variables, including age, family status, succession intentions, farm size, investment behaviour and fundamental values and motivations, on farmer behaviour and farm adjustment (MacFarlane, 1994). These filters condition farm response to external change, policy shifts in the context of this research, and thus interact to determine the spatial pattern of farm business and land-use change. Economically or environmentally deterministic models fail to represent this diversity of circumstance and outlook.

18.4   RESEARCH METHODOLOGY

A range of characteristics describing farmers as decision-makers, and details of their land holdings in the upland fringe of Grampian Region in the north-east of Scotland, were integrated within a GIS and used to predict overall farm response to defined policy scenarios. In essence this project attempted to represent the actual decision-making process on individual land holdings under conditions of external change. The difficulties, both technical and practical, in developing models of the land-use change process which represent the interaction of spatially variable environmental factors with farm-level data describing the resources and behavioural characteristics of individual holdings are considerable. This heterogeneity does, however, need to be represented to explore new approaches to policy impact modelling. This research has represented the conditioning effect of individuals' behavioural characteristics, in a highly abstracted form, as one component of a farm-change model using the ARC/INFO GIS.

A questionnaire survey of farms in the upland fringe of Grampian Region gathered details of 342 farms and their farmers. The data, describing the farm as a productive unit and as a business, together with the human, social and behavioural characteristics of farmers were spatially referenced to the point location of the farm buildings. A coverage of points was generated in ARC/INFO and polygons were buffered around each point, proportional in size to the area of that farm.

For each farm the interaction of three sets of data, illustrated in Figure 18.3, determined the outcome of policy shifts for each sampled farm holding. The questionnaire data were aggregated and classified into approximately 15 key characteristics, describing, for instance, the general economic status of the business, social stability of the farm household and the disposition of the farmer to a range of Alternative Farm Enterprises (AFEs).

Each sampled holding was classified by farm type, a classification used by government agriculture ministries. For each farm type, standard data describing the

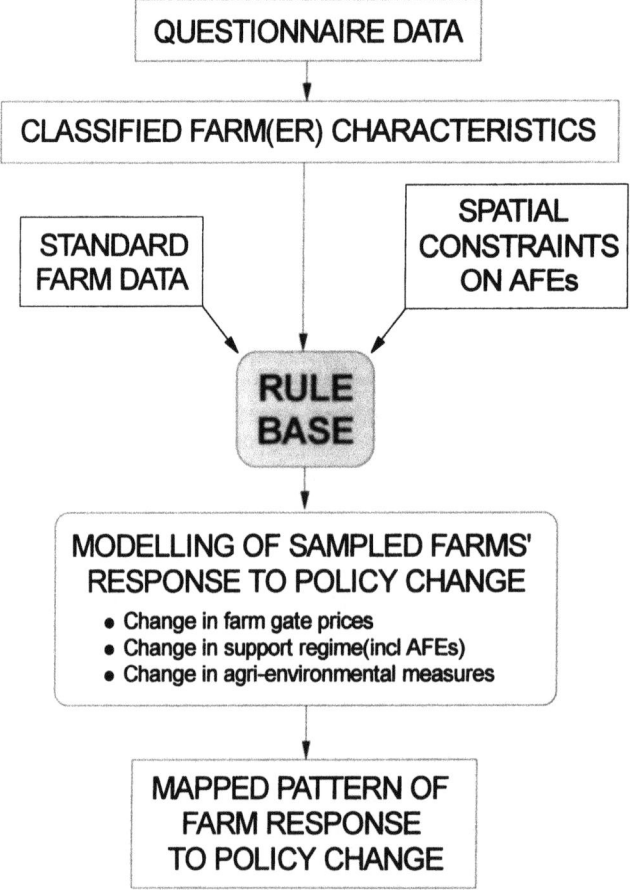

Figure 18.3 Data sources and rules in the farm change model.

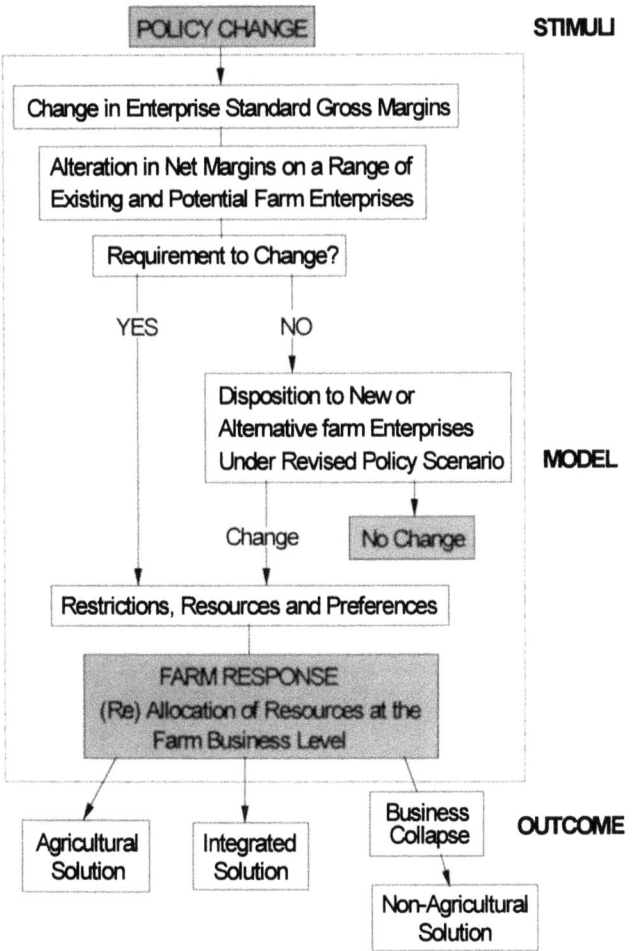

Figure 18.4   The farm-change model.

contribution of different enterprises such as crops, sheep and cattle to the overall farm output is available (Scottish Agricultural College, 1993), and this was used to translate the price changes for each scenario into farm-specific economic consequences. Clearly a scenario which massively reduced the value of sheep, but had limited effect on cereal prices, would affect a farm producing more sheep than anything else more than it would a specialist cereal producer. Standard data was thus used to quantify the differential impact of various scenarios.

The scenarios were not only concerned with agricultural production measures, but also with the potential development of AFEs such as tourist accommodation, 'paintball' or fishing. Research has shown that the successful development of many AFEs is related to accessibility and the amenity value of a farm's location (Ilbery, 1987). These factors were incorporated into the farm-change model as spatial datasets, utilising buffer zones from A-roads where accessibility was judged to be a significant factor, and designated areas of landscape importance (National Scenic Areas, for instance) in assessing the potential viability of farm-based holidays.

A rule-base was constructed to model the financial implications of each policy scenario for each farm and the potential for that business to survive into the short,

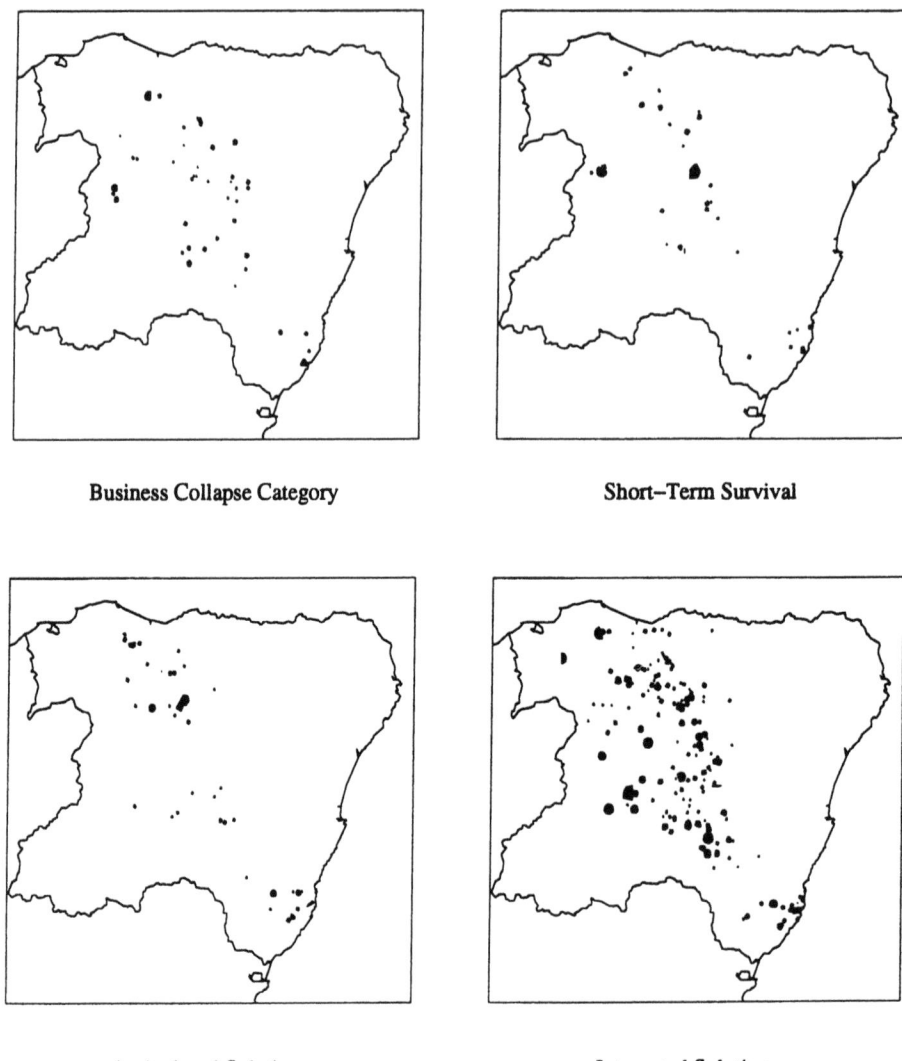

Business Collapse Category   Short–Term Survival

Agricultural Solution   Integrated Solution

Figure 18.5   Spatial distribution of farms by response category under a CAP reform scenario.

medium or long term. The rule base was structured such that the disposition of the farmer, and potential of the holding and business to successfully develop AFEs was of considerable significance in determining the farm-level outcome of the policy scenarios. Figure 18.4 illustrates the farm-change model as a decision tree, representing the sequential management decisions which farmers would need to make in response to a shift in their external production environment. For each of the farm polygons mapped, the model predicted the ability of the farm business to withstand the farm-gate price reductions consequent on policy change, and the subsequent adjustment of the farm to the new external environment. On the basis of the resources and viability of the farm business, constraints on the freedom of the farmer to pursue certain options (imposed, for instance, by tenancy restrictions) and the disposition of each individual decision-maker to alternative sources of income, sampled holdings were allocated to one of four categories: immediate business collapse, imminent business collapse (short-term survival), agricultural solution (continuing in non-diversified, viable, full-time farming) or integrated solution (continuing in farming, but supported by new or existing sources of non-agricultural/non-conventional income).

GIS provides the technology for the management and integration of aspatial and spatial data sets within a single database management system. ARC/INFO was used to integrate data describing the economic, social, human and behavioural characteristics of farmers and farms in the study area, with spatial data sets such as accessibility, land capability and composite data sets describing access to off-farm employment and the potential for the successful development of Alternative Farm Enterprises (AFEs).

18.5 MODEL OUTPUT

Figure 18.5 illustrates the spatial distribution of farms categorised into the four classes described for a relatively limited policy change scenario, assuming uncompensated falls in crop and livestock output value of between 10 per cent and 20 per cent. No spatial pattern was discernible for each of the farm response classes identified. The distribution of land managed by farms in each response class was analysed for spatial correlation with Land Capability for Agriculture (LCA) classes. No significant difference was observed between the distribution of all land managed by sampled farms across LCA classes, and the distribution of land managed by farms in each of the response categories. It is concluded therefore that the *process* of farm response to policy change is *primarily* a process with a spatial dimension and not an explicitly spatial process. The opportunities and constraints experienced by individual farmers are spatially variable, but the manner in which they are considered and utilised in farm adjustment is predominantly a reflection of the abilities, resources, awareness and freedom to act of individuals. Farm change is a highly complex function of spatially and temporally variable factors, combined with the personal, family and business characteristics of individual farmers and land managers.

18.6 CONCLUDING REMARKS

The critique presented here of previous, highly deterministic models of land-use change hinges on the concepts of scale and generalisation. The behavioural characteristics of individual farm-units were recognised in previous research to be a highly complex 'layer' in explaining and representing a system which determined the pattern of land-use and the spatially variable processes of change. Generalisations held this layer (which can be conceptualised as a spatially variable 'behavioural surface') as constant. Although the data requirements of constructing a comprehensive behavioural surface are considerable and problematic (Sorensen, 1992) this research has linked spatial, behavioural and economic elements which combine to condition farm-level response to policy change. The focus has been on individual farms, and the implications of change at the individual farm level. The research has developed a GIS-based tool to assess the implications of various policy changes on

a specific area of land, although the spatial coverage obtained though a questionnaire survey was based upon a relatively limited sample. The data requirements of
such an approach do represent an obstacle, but the validity of only pursuing applications of GIS for which there is adequate data (Openshaw, 1991) must be seriously
questioned in contexts where human behaviour is known to be a very significant
factor in the understanding and accurate representation of change.

> A GIS geography implies a neglect of the data ... GIS does not operate through
> connections derived from systems thinking; rather it treats data as discrete entities for
> which empirical relations can be sought. But data are outcomes of processes and
> mechanisms that can only be understood through some knowledge of how the data fit
> together as social organisation (which means a causal and not a mechanistic appre-
> ciation of the processes) (Taylor and Johnston, 1995, p. 59).

The study has shown that it is possible to construct more complex models of
landowner and land-user behaviour which will yield a more accurate picture of the
consequences of shifts in decision-makers' external environments. Such information
provides an essential input to the policy formulation debate, and the methodology
enhances the potential contribution of GIS in supporting future policy decisions,
including future research to extend the application from farm business response to
detailed changes in land-use.

REFERENCES

ASPINALL, R. (1991) Land-use impacts of climate change: an integrated research programme for Scotland. Paper presented at the Canada Centre for Remote Sensing, August
1991.
BARNES, T.J. (1988) Rationality and relativism in economic geography: an interpretative
review of the homo economicus assumption, *Prog. Hum. Geogr.* **12**, 473–496.
BROTHERTON, I. (1989) Farmer participation in voluntary land diversion schemes: some
observations from theory, *J. Rural Stud.* **5**, 299–304.
BROTHERTON, I. (1992) The success of set-aside and similar schemes. In GILG, A. W. ed.,
Restructuring the Countryside: Environmental Policy in Practice, Aldershot: Avebury.
CENTRE FOR AGRICULTURAL STRATEGY (1986) *Countryside Implications for England and
Wales of Possible Changes in the CAP*, (Main report). Reading: CAS.
COLMAN, D. (1994) Ethics and externalities: agricultural stewardship and other behaviour.
Presidential address to the Agricultural Economics Society Annual Conference, Exeter
University, April 1994.
ECONOMIST (1994) Rational economic man: the human factor. December 24th, 1994.
GILG, A. W. (1991) *Countryside Planning Policies for the 1990s*. Wallingford: CAB Interna-
tional.
HARVEY, D. R. (1992) Modelling land-use change: socio-economic factors. In WHITBY, M.
ed., *Land-use Change: Causes and Consequences*, Proceedings of a conference, Newcastle
upon Tyne, NERC. London: HMSO.
ILBERY, B. W. (1987) Geographical research into farm diversification: lessons for the exten-
sification proposals, Paper thirteen. In JENKINS, N. R. and BELL, B. eds, *Farm Exten-
sification: Implications of EC Regulation 1760/87*. Proceedings of a Workshop,
Merlewood Research and Development Paper No. 112, Institute of Terrestrial Ecology,
Grange-over Sands.
JOHNSTON, R. J. (1983) *Geography and Geographers*, 2nd edn. London: Edward Arnold.
MACFARLANE, R. (1994) Reductionist and deterministic approaches to farm adjustment
and land-use change: converging perspectives. Occasional Paper (New Series) No. 8,

Division of Geography and Environmental Management, University of Northumbria, Newcastle upon Tyne.

MACMILLAN, B. (ed.) (1989) *Remodelling Geography*. Oxford: Blackwell.

OPENSHAW, S. (1991) A view on the GIS crisis in geography, or, using GIS to put Humpty Dumpty back together again. *Environ. Plan.* A, **23**, 621–628.

POTTER, C. (1986) Processes of countryside change in lowland England. *J. Rural Stud.* **2**, 187–195.

POTTER, C. and GASSON, R. (1988) Farmer participation in voluntary land diversion schemes: some predictions from a survey. *J. Rural Stud.* **4**, 365–375.

SCOTTISH AGRICULTURAL COLLEGE (1993) *Farm Management Handbook 1993/4*. Edinburgh: SAC.

SORENSEN, E. M. (1992) Monitoring and managing rural changes – towards a socio-ecological methodology. *European GIS Conference 1992 Proceedings*, 364–371.

TARRANT, J. (1974) *Agricultural Geography*, Newton Abbot: David and Charles.

TAYLOR, P. J. and JOHNSTON, R. J. (1995) Geographic Information Systems and Geography. In PICKLES ed., *Ground Truth: the Social Implications of Geographic Information Systems*. New York: Guilford Press.

WHITE, B. (1990) The economic component of the NELUP project. NERC/ESRC Land-Use Programme Technical Report No. 12, University of Newcastle upon Tyne.

Index

9 780748 404599